Alakananda Bhattacharya
Amit Konar
Ajit K. Mandal

Parallel and Distributed Logic Programming

Towards the Design of a Framework
for the Next Generation Database
Machines

With 121 Figures and 10 Tables

 Springer

Dr. Alakananda Bhattacharya
Artificial Intelligence Laboratory
ETCE Department
Jadavpur University
Calcutta 700032
India
E-mail: b_alaka2@hotmail.com

Prof. Dr. Ajit K. Mandal
Artificial Intelligence Laboratory
ETCE Department
Jadavpur University
Calcutta 700032
India
E-mail: ajit.k.mandal@vsnl.com
        ajit.k.mandal@ieee.org

Prof. Dr. Amit Konar
Visiting Professor
Department of Math. and Computer
Science
University of Missouri, St. Louis
8001 Natural Bridge Road, St. Louis
Missouri 63121-4499
USA

Permanently working as
Professor
Department of Electronics and
Tele-communication Engineering
Jadavpur University
Calcutta 700032
India
E-mail: konaramit@yahoo.co.in

ISSN print edition: 1860-949X
ISSN electronic edition: 1860-9503

ISBN 978-3-642-07008-2   e-ISBN 978-3-540-33459-0

Springer is a part of Springer Science+Business Media
springer.com
© Springer-Verlag Berlin Heidelberg 2006
Softcover reprint of the hardcover 1st edition 2006

Cover design: deblik, Berlin

Preface

Foundation of logic historically dates back to the times of Aristotle, who pioneered the concept of truth/falsehood paradigm in reasoning. Mathematical logic of propositions and predicates, which are based on the classical models of Aristotle, underwent a dramatic evolution during the last 50 years for its increasing applications in automated reasoning on digital computers.

The subject of Logic Programming is concerned with automated reasoning with facts and knowledge to answer a user's query following the syntax and semantics of the logic of propositions/predicates. The credit of automated reasoning by logic programs goes to Professor Robinson for his well-known resolution theorem that provides a general scheme to select two program clauses for deriving an inference. Until now Robinson's theorem is being used in PROLOG/DATALOG compilers to automatically build a Select Linear Definite (SLD) clause based resolution tree for answering a user's query.

The SLD-tree based scheme for reasoning undoubtedly opened a new era in logic programming for its simplicity in implementation in the compilers. In fact, SLD-tree construction suffices the need for users with a limited set of program clauses. But with increase in the number of program clauses, the execution time of the program also increases linearly by the SLD-tree based approach. An inspection of a large number of logic programs, however, reveals that more than one pair of program clauses can be resolved simultaneously without violating the syntax and the semantics of logic programming. This book employs this principle to speed up the execution time of logic programs.

One question that naturally arises: how does one select the clauses for concurrent resolution? Another question that crops up in this context: should one select more than two clauses together or pairs of clauses as groups for concurrent resolution? This book answers these questions in sufficient details. In fact, in this book we minimize the execution time of a logic program by grouping sets of clauses that are concurrently resolvable. So, instead of pairs, groups of clauses with more than two members in a group are resolved at the same time. This may give rise to further questions: how can we ensure that the selected groups only are concurrently resolvable, and members in each group too are maximal? This in fact is a vital question as it ensures the optimal time efficiency (minimum execution time) of a logic program. The optimal time efficiency in our proposed system is attained by mapping the program clauses onto a specialized structure that allows each group of resolvable clauses to be mapped in close proximity, so as to participate in the resolution process. Thus n-groups of concurrently resolvable clauses form n clusters in the network. Classical models of Petri nets have been extended to support the aforementioned requirements.

Like classical Petri nets, the topology of network used in the present context is a bipartite graph having two types of nodes, called places and transitions, and directed arcs connected from places to transitions and transitions to places respectively. Clauses describing IF-THEN rules (knowledge) are mapped at the transitions, with predicates in IF and THEN parts being mapped at the input and the output places of the transitions. Facts described by atomic predicates are mapped at the places that too share predicates of the IF or the THEN parts of a rule. As an example, let us consider a rule: (Fly(X) ¬Bird(X).) and a fact: (Bird(parrot)¬.). The above rule in our terminology is represented by a transition with one input and one output place. The input and the output places correspond to the predicates: Bird(X) and Fly(X) respectively. The fact: Bird(parrot) is also mapped at the input place of the transition. Thus, a resolution of the rule and the fact is possible because of their physical proximity on the Petri net architecture. It can be proved by method of induction easily that all members in a group of resolvable clauses are always mapped on the Petri net around a transition. Thus a number of groups of resolvable clauses are mapped on different transitions and the input-output places around them. Consequently, a properly designed firing rule can ensure concurrent resolution of the groups of clauses and generation and storage of the inferences at appropriate places. The book aimed at realizing the above principle by determining appropriate control signals for transition firing and resulting token saving at desired places.

It is indeed important to note that the proposed scheme of reasoning covers the notion of AND-, OR-, Stream- and Unification-parallelisms. It is noteworthy that there are plenty of research papers with hundreds of scientific jargons to prohibit the unwanted bindings in AND-parallelisms, but very few of them are realistic. Implementation of the Stream-parallelism too is difficult, as it demands design of complex control strategies. Fortunately, because of the structural benefits of Petri nets, AND- and Stream-parallelisms could have been realized by our proposed scheme of concurrent resolution automatically. The most interesting point to note is that these parallelisms are realized as a byproduct of the adopted concurrent resolution policy, and no additional computation is needed to implement the former.

The most important aspect of this book, probably, is the complete realization of the proposed scheme for concurrent resolution on a massively parallel architecture. We verified the architectural design with VHDL and the implementations were found promising. The VHDL source code is not included in the book for its sheer length that might have enhanced its volume three times its current size. Finally, the book concludes on the possible application of the proposed parallel and distributed logic programming for the next generation database machines.

Studies in Computational Intelligence, Volume 24

Editor-in-chief
Prof. Janusz Kacprzyk
Systems Research Institute
Polish Academy of Sciences
ul. Newelska 6
01-447 Warsaw
Poland
E-mail: kacprzyk@ibspan.waw.pl

Further volumes of this series
can be found on our homepage:
springer.com

Vol. 8. Srikanta Patnaik, Lakhmi C. Jain,
Spyros G. Tzafestas, Germano Resconi,
Amit Konar (Eds.)
Innovations in Robot Mobility and Control,
2005
ISBN 3-540-26892-8

Vol. 9. Tsau Young Lin, Setsuo Ohsuga,
Churn-Jung Liau, Xiaohua Hu (Eds.)
*Foundations and Novel Approaches in Data
Mining,* 2005
ISBN 3-540-28315-3

Vol. 10. Andrzej P. Wierzbicki, Yoshiteru
Nakamori
Creative Space, 2005
ISBN 3-540-28458-3

Vol. 11. Antoni Ligęza
*Logical Foundations for Rule-Based
Systems,* 2006
ISBN 3-540-29117-2

Vol. 12. Jonathan Lawry
*Modelling and Reasoning with Vague Con-
cepts,* 2006
ISBN 0-387-29056-7

Vol. 13. Nadia Nedjah, Ajith Abraham,
Luiza de Macedo Mourelle (Eds.)
Genetic Systems Programming, 2006
ISBN 3-540-29849-5

Vol. 14. Spiros Sirmakessis (Ed.)
Adaptive and Personalized Semantic Web, 2006
ISBN 3-540-30605-6

Vol. 15. Lei Zhi Chen, Sing Kiong Nguang,
Xiao Dong Chen
*Modelling and Optimization of
Biotechnological Processes,* 2006
ISBN 3-540-30634-X

Vol. 16. Yaochu Jin (Ed.)
Multi-Objective Machine Learning, 2006
ISBN 3-540-30676-5

Vol. 17. Te-Ming Huang, Vojislav Kecman,
Ivica Kopriva
*Kernel Based Algorithms for Mining Huge
Data Sets,* 2006
ISBN 3-540-31681-7

Vol. 18. Chang Wook Ahn
Advances in Evolutionary Algorithms, 2006
ISBN 3-540-31758-9

Vol. 19. Ajita Ichalkaranje, Nikhil
Ichalkaranje, Lakhmi C. Jain (Eds.)
*Intelligent Paradigms for Assistive and
Preventive Healthcare,* 2006
ISBN 3-540-31762-7

Vol. 20. Wojciech Penczek, Agata Półrola
*Advances in Verification of Time Petri Nets
and Timed Automata,* 2006
ISBN 3-540-32869-6

Vol. 21. Cândida Ferreira
*Gene Expression on Programming: Mathematical
Modeling by an Artificial Intelligence,* 2006
ISBN 3-540-32796-7

Vol. 22. N. Nedjah, E. Alba, L. de Macedo
Mourelle (Eds.)
Parallel Evolutionary Computations, 2006
ISBN 3-540-32837-8

Vol. 23. M. Last, Z. Volkovich, A. Kandel (Eds.)
Algorithmic Techniques for Data Mining, 2006
ISBN 3-540-33880-2

Vol. 24. Alakananda Bhattacharya, Amit Konar,
Ajit K. Mandal
Parallel and Distributed Logic Programming,
2006
ISBN 3-540-33458-0

Alakananda Bhattacharya, Amit Konar, Ajit K. Mandal
Parallel and Distributed Logic Programming

1

An Introduction to Logic Programming

This chapter provides an introduction to logic programming. It reviews the classical logic of propositions and predicates, and illustrates the role of the resolution principle in the process of execution of a logic program using a stack. Local stability analysis of the interpretations in a logic program is undertaken using the well-known "s-norm" operator. Principles of "data and instruction flow" through different types of parallel computing machines, including SIMD, MIMD and data flow architectures, are briefly introduced. Possible parallelisms in a logic program, including AND-, OR-, Stream- and Unification-parallelisms, are reviewed with an ultimate aim to explore the scope of Petri net models in handling the above parallelisms in a logic program.

1.1 Evolution of Reasoning Paradigms in Artificial Intelligence

The early traces of Artificial Intelligence are observed in some well-known programs of game playing and theorem proving of the 1950's. The *Logic Theorist* program by Newell and Simon [26] and the *Chess playing* program by Shannon [33] need special mention. The most challenging task of these programs is to generate the state-space of problems by a limited number of rules, so as to avoid the scope of combinatorial explosion. Because of this special characteristic of these programs, McCarthy coined the name Artificial Intelligence [2] to describe programs showing traces of intelligence in determining the direction of moves in a state-space towards the goal.

Reasoning in early 1960's was primarily accomplished with the tools and techniques of production systems. The DENDRAL [4] and the MYCIN [35] are the two best-known and most successful programs of that time, which were designed using the formalisms of production systems. The need for logic in building intelligent reasoning programs was realized in early 1970's. Gradually, the well-known principles of propositional and predicate logic were reformed for applications in programs with more powerful reasoning capability. The most successful program exploring logic for reasoning perhaps is MECHO. Designed by Bundy [5] in late 1970's, the MECHO program was written to solve a wide

A. Bhattacharya et al.: *An Introduction to Logic Programming*, Studies in Computational Intelligence (SCI) **24**, 1–55 (2006)
www.springerlink.com

range of problems in Newtonian mechanics. It uses the formalisms of meta-level inference to guide search over a range of different tasks, such as common sense reasoning, model building and the manipulation of algebraic expressions for equation solving [14]. The ceaseless urge for realizing human-like reasoning on machines brought about a further evolution in the traditional logic of predicates in late 1980s. A new variety of predicate logic, which too deals with the binary truth functionality of predicates but differs significantly from the reasoning point of view, emerged in the process of evolution. The fundamental difference in reasoning of the deviant variety of logic with the classical logic is that a reasoning program implemented with the former allows contradiction of the derived inferences with the supplied premises. This, however, is not supported in the classical logic. The new class of logic includes non-monotonic logic [1], default logic [3], auto-epistemic logic [23], modal logic [30] and multi-valued logic [16].

In late 1980's, a massive change in database technology was observed with the increasing use of computers in office automation. Commercial database packages, which at that time solely rested on hierarchical (tree-structured) and network models of data, were fraught with the increasing computational impacts of the relational paradigms. The relational model reigned the dynasty of database systems for around a decade, but gradually its limitations too in representing complex *integrity constraints* were shortly discovered. To overcome the limitations of the relational paradigms, the database researchers took active interest in employing logic to model database systems. Within a short span of time, one database package, called Datalog, that utilizes the composite benefits of relational model and classical logic emerged. The Datalog programs are similar to PROLOG programs that answer a user's query by a depth-first traversal over the program clauses. Further, for satisfaction of a complex goal, that includes conjunction of several predicates, the Datalog program needs to backtrack to the previous program clauses. Unfortunately, the commercial work stations/main frame machines that usually offers array as their elementary data structure are inefficient to run Datalog programs that requires tree/stack as the primary program resources.

To facilitate the database machines with the computational power of efficiently running Datalog programs, a significant amount of research was undertaken in various research institutes of the globe since 1990. Some research groups emphasized the scope of parallelism in runtime [30, 36, 38] of a Datalog program, some considered the scope of resolving parallelism in the compile-time phase [10, 37], and the rest took interest to model parallelism in the analysis phase [34]. However, no concrete solution to the problem was reported till this date. The book attempts a new approach to design a parallel architecture for a Datalog-like program, which is capable of overcoming all the above limitations of the last 30 years' research on *logic program based machines*.

1.2 The Logic of Propositions and Predicates — A Brief Review

The word 'proposition' stands for a fact, having a binary valuation space of {true, false}. Thus a fact, that can be categorized to be true or false, is a proposition. Since the beginning of the last century philosophers have devised several methods to determine the truth or falsehood of an inference [27] from a given set of facts. The process of deriving the truth-value of a proposition from the known truth-values of its premises is called *reasoning* [20]. Both semantic and syntactic approaches to reasoning are prevalent in the current literature of Artificial Intelligence. The semantic approach [11] employs a truth table for estimation of the truth-value of a rule from its premise clauses. When a rule depends on n number of premise clauses, the number of rows in the table becomes as large as 2^n. The truth table approach thus has its inherent limitation in reasoning applications.

The syntactic approach on the other hand employs syntactic rules to logically derive the truth-value of a given clause from the given premises. One simple syntactic rule, for instance, is the chain rule, given below:

$$Chain\ rule:\ p \rightarrow q,\ q \rightarrow r \Rightarrow p \rightarrow r. \tag{1.1}$$

where p, q and r are atomic propositions, '$\rightarrow$' denotes an if-then operator and '$\Rightarrow$' denotes an implication function.

The linguistic explanation of the above rule is "given: *if p then q* and *if q then r*, we can then infer *if p then r*". With such a rule and a given fact p, we can always infer r. Formally,

$$p,\ p \rightarrow q,\ q \rightarrow r \Rightarrow r. \tag{1.2}$$

The statement (1.2) is an example of inferencing by a syntactic approach. In fact, there exists around 20 rules like the chain rule, and one can employ them in a reasoning program to determine the truth-value of an unknown fact. A complete listing of these rules is available in any standard textbook on AI [15].

Propositional logic was well accepted, both in the disciplines of Philosophy and Computer Science. But shortly its limitations in representing complex real world knowledge became pronounced. Two major limitations of propositional logic are (i) *incapability of representing facts with variables as arguments* and (ii) *lack of expressing power of quantifiers like 'for all ($\forall$)' and 'for some ($\exists$)'*. These two limitations led the researchers to extend the syntactical power of propositional logic. The logic that came up shortly free from these limitations is called *'the logic of predicates'* or *'predicate logic'* in brief. The following statements illustrate the power of expressing complex statements by predicate logic.

Statement 1: *All boys like flying kites.*

   Representation in predicate logic:

   $\forall X$ (Boy (X)$\rightarrow$ Likes (X, flying -kites)).   (1.3)

Statement 2: *Some boys like sweets.*

   Representation in predicate logic:

   $\exists X$ (Boy (X)$\rightarrow$ Likes (X, sweets)).   (1.4)

   In the last two statements, we have predicates like Boy and Likes that have a valuation space of {true, false} and terms like X, sweets, and flying-kites. In general, a *term* can be a variable like X or a constant like sweets or flying-kites or even a function or function of function (of variables). The next example illustrates functions as terms in the argument of a predicate.

Statement 3: *For all X if* $(f (X) > g (X))$
                  then $(f (g (X)) = g (f (X))).$

Representation in predicate logic:

   $\forall X$ (Greater-than (f (X), g (X))$\rightarrow$
   Equal (f (g (X)), g (f (X)))).   (1.5)

   The last statement in predicate logic been self-explanatory, needs no further explanation.
   Given a set of facts and rules (piece of knowledge), we can easily derive the truth or falsehood of a predicate, or evaluate the value of the variables used in the argument of predicates. The process of evaluation of the variables or testing the truth or falsehood of predicates is usually called '*inferencing*' [32]. There exists quite a large number of well-known inferential procedures in predicate logic. The most common among them is the 'Robinson's inference rule', popularly known as the '*resolution principle*'. The resolution principle is applicable onto program clauses expressed in Conjunctive Normal Forms (CNF).
   Informally, a CNF of a clause includes disjunction (OR) of negated or non-negated literals. A general clause that includes conjunction of two or more CNF sub-clauses is thus re-written as a collection of several CNF sub-clauses.
   For example the following two program clauses, containing literals P_{ij} and Q_{ij} for $1 \le i \le n$ and $1 \le j \le m$, are expressed in CNF.

$$\left.\begin{array}{l} \neg P_{11} \vee \neg P_{12} \vee .....\vee \neg P_{1n} \vee Q_{11} \vee Q_{12} \vee .....\vee Q_{1m}. \\[2ex] \neg P_{21} \vee \neg P_{22} \vee .....\vee \neg P_{2n} \vee Q_{21} \vee Q_{22} \vee ......\vee Q_{2m}. \end{array}\right\} \qquad (1.6)$$

It may be noted from statement (1.6) above that program clauses expressed in CNF are free from conjunction (AND) operators. The principle of resolution of two clauses expressed in CNF is now outlined.

1.2.1 The Resolution Principle

Consider predicates P, Q_1, Q_2 and R. Let us assume that with appropriate substitution S,

$Q_1 [S] = Q_2 [S]$.

Then $(P \vee Q_1) \wedge (\neg Q_2 \vee R)$ with $Q_1 [S] = Q_2 [S]$ yields $(P \vee R) [S]$.

$$\text{Symbolically,} \quad \frac{P \vee Q_1, \neg Q_2 \vee R \quad\quad Q_1 [S] = Q_2 [S]}{(P \vee R) [S]} \tag{1.7}$$

Example 1.1 illustrates the resolution principle.

Example 1.1: Let P = Loves (X, Father-of (X)),

$$\left. \begin{array}{l} Q_1 = \text{Likes (X, Mother-of (X))}, \\ Q_2 = \text{Likes (john, Y)}, \\ R = \text{Hates (X, Y)}. \end{array} \right\} \tag{1.8}$$

After unifying Q_1 and Q_2, we have

$Q = Q_1 = Q_2$ = Likes (john, Mother-of (john)).

where the substitution S is given by

S = {john/X, Mother-of (X)/Y}
= {john/X, Mother-of (john)/Y}.

The resolvent $(P \vee R) [S]$ is, thus, computed as follows:

$(P \vee R) [S]$ =
Loves (john, Father-of (john)) $\vee$ Hates (john, Mother-of (john)).

The substitution S in many books is denoted by s and Q [S] is denoted by Qs. In fact, we shall adopt the latter notion later in this book.

1.2.2 Theorem Proving in the Classical Logic with the Resolution Principle

Suppose, we have to prove a theorem Th from a set of axioms. We denote it by

{ A_1, A_2, ..., A_n} $\models$ Th

Let
    A_1 = Biscuit (coconut-crunchy)
    A_2 = Child (mary) $\wedge$ Takes (mary, coconut-crunchy)
    A_3 = $\forall$ X (Child(X) $\wedge$ $\exists$ Y (Takes (X,Y) $\wedge$ Biscuit (Y)))
        $\rightarrow$ Loves (john, X)

(1.9)

and
    Th = Loves (john, mary) = A_4 (say).

Now, to prove the above theorem, we first express clauses A_1 through A_4 in CNF. Expressions A_1 and A_4 are already in CNF. Expression A_2 can be converted into CNF by breaking it into two clauses:

        Child (mary)
and   Takes (mary, coconut-crunchy).

Further, the CNF of expression A_3 is

$\neg$Child (X) $\vee$ $\neg$Takes (X,Y) $\vee$ $\neg$Biscuit (Y) $\vee$ Loves (john, X)

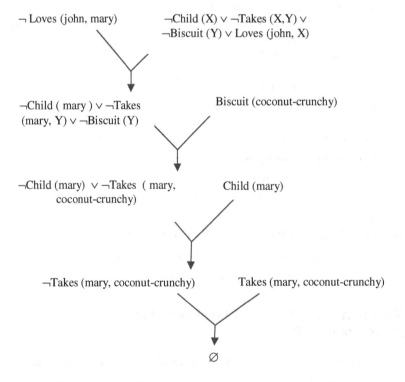

Fig. 1.1: A resolution tree constructed to prove that Loves (john, mary)

Now it can be easily shown that the negation of the theorem (goal) if resolved with the CNF form of expressions A_1 through A_3, the resulting expression would be a null clause for a valid theorem. To illustrate this, we will now form pairs of clauses, one of which contains a positive predicate, while the other contains the same predicate in negated form. Thus by the resolution principle, both the negated and positive literals will drop out and the value of the variables used for unification should be substituted in the resulting expression. The principle of resolution is illustrated in Fig. 1.1 to prove the goal that Loves (john, mary).

The resolution principle has a logical basis, and a mathematical proof of its soundness and completeness is also available in [2]. We instead of proving these issues once again, just state their definitions only.

Definition 1.1: *The resolution theorem is **sound** if any inference α that has been proved from a set of axioms S by the resolution theorem, i.e., $S \vdash \alpha$, we can show that α logically follows from S, by notation, $S \models \alpha$.*

Definition 1.2: *The resolution theorem is called **complete**, if for any inference α, that follows logically from S, i.e., $S \models \alpha$, we can prove by the resolution theorem $S \vdash \alpha$.*

Because of the aforementioned two characteristics of the resolution theorem, it found a wide acceptance in automating the inferencing process in predicate logic.

1.3 Logic Programming

The statements in predicate logic have the following form in general

$Q (P_1 \text{ (arguments) } \wedge P_2 \text{ (arguments) } \wedge \ldots \ldots \wedge P_n \text{ (arguments)} \rightarrow$
$(Q_1 \text{ (arguments) } \vee Q_2 \text{ (arguments) } \vee \ldots \ldots \vee Q_m \text{ (arguments)}))$. (1.10)

where Q is the quantifier ($\forall, \exists$), P_i and Q_j are predicates. It is to be noted that the above rule includes a number of 'V' operators in the right-hand side of the '$\rightarrow$' operator. Since the pre-condition of any Q_j are all the P_i s , we can easily write the above expression in CNF form as follows.

$Q (P_1 \text{ (arguments) } \wedge P_2 \text{ (arguments) } \wedge \ldots \ldots \wedge P_n \text{ (arguments)}$
$\rightarrow Q_1(\text{arguments}))$.

$Q (P_1 \text{ (arguments) } \wedge P_2 \text{ (arguments) } \wedge \ldots \ldots \wedge P_n \text{ (arguments)}$
$\rightarrow Q_2(\text{arguments}))$. (1.11)

$Q (P_1 \text{ (arguments) } \wedge P_2 \text{ (arguments) } \wedge \ldots \ldots \wedge P_n \text{ (arguments)}$
$\rightarrow Q_m \text{ (arguments)}))$.

In such a representation there exists only one predicate in the then part (consequent part) of each clause. Such representation of clauses, where the then part contains at most one literal, is the basis of logic programs.

1.3.1 Definitions

The definitions[1] that best describe a logic program are presented below in order.

Definition 1.3: *A **horn clause** is a clause with at most one literal in the then part (head) of the clause.* For instance

$$P (X, Y) \leftarrow Q (Y, X). \hspace{5cm} (1.12)$$
$$P (X, Y) \leftarrow Q (Y, X), R (X, Z), S (Z). \hspace{2.5cm} (1.13)$$
$$P (a, b) \leftarrow. \hspace{6.5cm} (1.14)$$
$$\leftarrow Q (Y, X). \hspace{6.5cm} (1.15)$$

are some of the typical example of horn clauses. It is to be noted that in the clauses (1.12) to (1.15), (1.12) and (1.13) are rules, (1.14) is a fact and (1.15) is a query.

Definition 1.4: *A **logic program** is a collection of horn clause statements.*

An example of a typical logic program with a query is in example 1.2.

Example 1.2: The clauses listed under (1.16) describe a typical logic program and clause (1.17) denotes its corresponding query.

$$\left. \begin{array}{l} \text{Can-fly (X)} \leftarrow \text{Bird (X), Has-wings (X).} \\ \text{Bird (parrot)} \leftarrow. \\ \text{Has-wings (parrot)} \leftarrow. \end{array} \right\} \hspace{2cm} (1.16)$$

$$\text{Query: } \leftarrow \text{Can-fly (parrot).} \hspace{3cm} (1.17)$$

Definition 1.5: *When there exists one literal in the heads of all the clauses, the clauses are called **definite**, and the corresponding logic program is called a **definite program**.*

The logic program given in example 1.2 is a definite program as all its constituent clauses are definite.

---

[1] These definitions are formally given once again in chapter 3 for the sake of completeness.

1.3.2 Evaluation of Queries with a Stack

Given a logic program and a user's query. A resolution tree is gradually built up and traversed in a depth first manner to answer the user's query. For realization of the depth first traversal on a tree we require a stack. The principle of the tree building process and its traversal is introduced here with an example presented later. The stack to be employed for the tree construction has two fields, one field containing the orderly traversed nodes (resolvents) of the tree, and the other field holds the current set of variable bindings needed for generating the resolvents.

Like conventional stacks the stack pointer (top) here also points to the top of the stack, up to which the stack is filled in. Initially, the query is pushed into the stack. Since it has no variable bindings until now, the variable bindings' field is empty. The query clause may now be resolved with one suitable clause from the given program clauses, and the resulting clause and the variable bindings used to generate it are then pushed into the stack. Thus as the tree is traversed downward a new node describing a new resolvent is created and pushed into the stack. The process of pushing into the stack continues until a node is reached which either yields a null clause, or cannot be resolved with any available program clause. Such nodes are called *leaves/dead ends* of the tree. Under this circumstance, we may require to move to the parent of a leaf node to take a look for an alternative exploration of the search space. The moving up process in the tree is accomplished by popping the stack out. The popped out node denotes the parent of the current leaf node. The process of alternative resolution with the popped out node is then examined and the expansion of the tree is continued until the root node in the tree is reached again. Example 1.3 illustrates resolution by stack.

Example 1.3: Resolution by Stack

Logic Program:

$$
\left.\begin{array}{ll}
1. & P(X, Y) \leftarrow Q(Y, Z), R(Z). \\
2. & R(C) \leftarrow. \\
3. & Q(b, b) \leftarrow. \\
4. & R(b) \leftarrow.
\end{array}\right\} \qquad (1.18)
$$

A traversal on the tree for answering the query: $\leftarrow P(a, b)$ is presented in Fig. 1.2. When a node is expanded by resolution the child of the said node is pushed into the Stack Pointer (SP) moves up one position to indicate the latest information in the stack. When a node cannot be expanded, it is popped out from the stack, and the next node in the stack is considered for possible expansion. The resolution tree is terminated when construction process of the stack top is filled with a null clause.

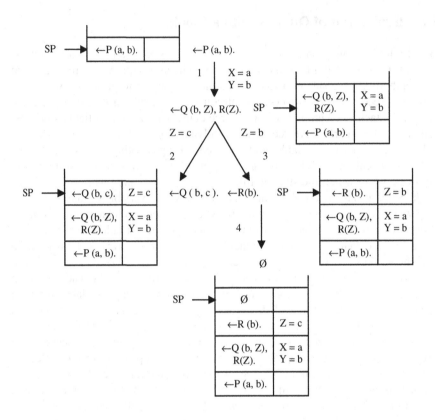

Fig. 1.2: Depth-first traversal on a tree to answer the user's query

1.3.3 PROLOG — An Overview

'PROLOG' is an acronym for PROgramming in LOGic. It is the most popular programming language for logic programming. The advantage of PROLOG over the conventional procedural programming languages like C or Pascal and the functional programming language like LISP is manifold. The most useful benefit that the programmer can derive from PROLOG is the simplicity in programming. Unlike the procedural languages, where the procedure for a given problem has to be explicitly specified in the program, a PROLOG program only defines the problem by facts and if-then rules, but does not include any procedure to solve the problem. In fact, the compiler of PROLOG takes the major role of automatically matching the part of one clause with another to execute the process of resolution. The execution of a PROLOG program thus is a sequence of steps of resolution over the clauses in the program, which is usually realized with a stack, as discussed in section 1.3.2. One step of resolution over two clauses thus calls for a PUSH operation on the stack. On failure (in the process of matching) the control POPs the stack return to the parent of the currently invoked clause. One most useful built-in predicate in PROLOG is 'CUT'. On failure it helps the control to return to the root of the 'resolution tree' for re-invoking the search and resolution process beginning from the root. This, however, has a serious drawback as a part of the 'resolution tree' (starting from the root where the resolution fails) remains unexplored. To avoid unwanted return to the root for subsequent search for resolution, the clause comprising of the CUT predicate has a special structure. For example, consider the following clause using propositions and CUT predicate only (for brevity).

$$Cl \leftarrow p, q, !, r, s \qquad\qquad (1.19)$$

where 'Cl' is the head of the given clause, p, q, ! (CUT), r and s are in the body. It is desired that if any proposition preceding CUT fails, the control returns to the parent of the present clause. But if all literals (p and q) preceding CUT are satisfied, then CUT is automatically satisfied. Consequently, if any literal like r or s fails, the control returns to the root of the resolution tree.

Further, unlike arrays in most procedural languages, tree is the basic data structure of PROLOG and depth first traversal is the built-in feature of PROLOG for clause invocation and resolution process.

The early versions of PROLOG compiler did not have the provision for concurrent invocation of the program clauses. The later version of PROLOG (for instance PARLOG [6]) includes the feature of concurrency in the resolution process. In this book, we will present some schemes for parallel realizations of logic programs in runtime.

One interesting point to note is that all the resolvents obtained through resolution principle may not be equally stable. Consequently a question of relative stability [25] appears in the interpretation of a logic program. The next section provides a brief introduction to determining stable interpretation of a logic program. A detailed discussion on this, which is available elsewhere ([7], [8]), is briefly outlined below for the sake of completeness of the book.

1.3.4 Interpretation and their Stability in a Logic Program

Usually a logic program consists of a set of Horn clauses. An interpretation of the logic program thus refers to the intersection of the interpretations of the individual clauses. Example 1.4 illustrates the aforementioned principles.

Example 1.4: Consider the following two clauses:

    1. $q \leftarrow p$.

    2. $p \leftarrow$.

We need to determine the common interpretation of the given clauses.

Let the interpretation of the clauses (1) and (2) be denoted by I_1 and I_2 respectively. Here, $I_1 = \{(p, d)\}$ where d denotes the don't care state of q, and $I_2 = \{(p, q), (\neg p, q), (\neg p, \neg q)\}$. Therefore, the common interpretation of two clauses is given by

$$I = I_1 \cap I_2$$
$$= \{(p, d)\} \cap \{(p, q), (\neg p, q), (\neg p, \neg q)\}$$
$$= \{(p, q)\},$$

signifying that p and q are both true.

The interpretation of the given logic program has been geometrically represented in Fig. 1.3.

An important aspect of logic programs that need special consideration: whether all interpretations of a given clause are equally stable? Some works on stability analysis of logic programs have already been reported in [1], [3] and [19]. Unfortunately the methodology of stability analysis applied to logic programs is different, and there is no unified notion of stability analysis until this date. On the other hand a lot of classical tools of cybernetic theory such as, energy minimization by Liapunov energy function (vide [18]), Routh-Hurwitz criterion (vide [17]), Nyquist criterion [28] etc. are readily available for determining stability of any complex nonlinear system. In recent times researchers are taking keen interest to use these classical theories in the stability analysis of logic programs as well [9]. In this section we briefly outline a principle of stability analysis by replacing AND-operator by t-norm and OR-operator by s-norm. It may be added here that the advantage of using these norms is to keep the function continuous and hence differentiable. Example 1.5 briefly outlines the principle of determining stable points in a logic program.

Example 1.5: We consider to determine the stable (or at least relatively more stable) interpretation of the clause 'q $\leftarrow$ p.'. Replacing 'q $\leftarrow$ p.' by '$\neg p \lor q$' and then further replacing 'OR ($\lor$)' by s-norm [16], where for any two propositions a and b,

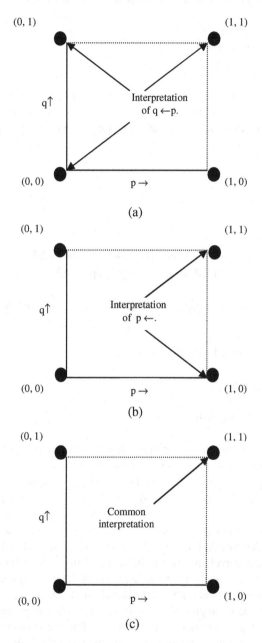

Fig. 1.3: Geometric representation of the common interpretation for the given logic program

a s b = a + b – ab, we can construct a relation F(p, q) as follows:

$$
\begin{aligned}
F(p, q) &= (1 - p)\, sq \\
&= (1 - p) + q - (1 - p)\, q \\
&= 1 - p + pq.
\end{aligned}
$$

It can be verified that $F(p, q) = 1 - p + pq$ satisfies all the three interpretations of 'q ←p.' To determine the stable points, if any, on the constructed surface of F(p, q), let us presume that there exists at least one stable point (p^*, q^*). We now perturb (p^*, q^*) by (h, δ), i.e.,

$$
\begin{aligned}
p &= p^* + h \\
q &= q^* + \delta
\end{aligned}
$$

Thus, we obtain:

$$
\begin{aligned}
F(p^* + h, q^* + \delta) &= 1 - p^* + p^*q^* - h + hq^* + \delta p^* + h\delta \\
&= F(p^*, q^*) - h + hq^* + \delta p^* + h\delta
\end{aligned} \tag{1.20}
$$

Now for stability of the given clause at (p^*, q^*), we need to satisfy the following condition:

$$
F(p^* + h, q^* + \delta) = F(p^*, q^*),
$$

which ultimately demands

$$
(-h + hq^* + \delta p^* + h\delta) = 0. \tag{1.21}
$$

It can be verified that the aforementioned condition is satisfied only at (p^*, q^*) = $(\neg p, q)$, irrespective of any small value of h and δ. However, if we put other two interpretations of 'q ←p.', such as (p, q), $(\neg p, \neg q)$ in condition (1.21), we note that it imposes restriction on h and δ, which are not feasible. Thus the interpretation $(\neg p, q)$ is a stable point. More interesting results on stability analysis are shortly to appear in a forthcoming paper [9] from our research team.

The problem of determining stability for non-monotonic and default logic is more complex. This, however, is beyond the scope of the present book. Very few literatures dealing with the analysis of stable points of default and non-monotonic logic are available in the current realm of Artificial Intelligence [22, 31].

In this book our main emphasis is on the design of a high speed parallel architecture for the logic programming machines. For the convenience of the readers we briefly outline the principles of parallelism and pipelining, and various configurations of parallel computing machines.

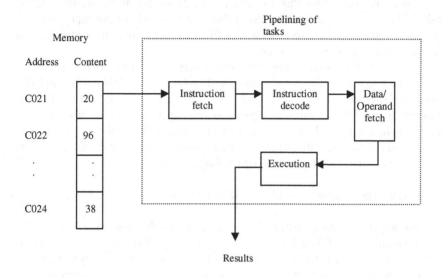

Fig. 1.4: The pipelining concept

1.4 Introduction to Parallel Architecture

Parallelism and pipelining are two important issues in high-speed computation of programs. Usually, these two concepts rest on the principles of *Von Neumann machines* [13], where the instructions are fetched from a given storage (memory) and subsequently executed by a hardwired engine called the *central processing unit* (CPU). The execution of an instruction in a program thus calls for four major steps: (i) instruction fetching, (ii) instruction decoding, (iii) data/operand fetching and (iv) activating the arithmetic and logic unit (ALU) for executing the instruction. These four operations are usually in pipeline (vide Fig. 1.4).

It needs mention that the units used in a pipelined system must have different tasks and each unit (except the first) should wait for another to produce the input for it.

Unlike pipelining, the concept of parallel processing calls for *processing elements* (PE) having similar tasks. A task allocator distributes the concurrent (parallel) tasks in the program and allocates them to the different processing elements. As an example, consider the program used for evaluation of Z where

$$Z = P * Q + R * S, \qquad (1.22)$$

where P, Q, R and S are real numbers.

Suppose we have two PEs to compute Z. Since the first part (P * Q) is independent from the second (R * S) in the right hand side of the expression, we can easily allocate these two tasks to two PEs, and the results thus obtained may be added by either of them to produce Z.

A schematic diagram depicting the above concept is presented in Fig. 1.5.

For the last figure after computation of (P * Q) and (R * S) by PE_1 and PE_2 respectively, either of the results ($Temp_1 = P * Q$ or $Temp_2 = R * S$) needs to be transferred to PE_1/PE_2 for the subsequent addition operation. So, the task allocator has to re-allocate the task of addition to either of PE_1 or PE_2 in the second cycle. It is undoubtedly clear that the two cycles are required to complete the task, which in absence of either of the PEs would require three cycles.

Thus the speed-up factor = $(2/3) \times 100 = 66.66\%$.

Generally, a switching network is connected among the processing elements for the communication of data from one PE to the others. Depending on the type of concurrency among the tasks, parts of the switching network needs to be activated in time sequence. Among the typical switching networks cubes, barrel shifters, systolic arrays, etc. need special mention.

Depending on the flow of instructions and data among the processing elements and memory units, four different topologies of machines are of common interest to the professionals of computer architecture. These machines are popularly known as Single Instruction Single Data (SISD), Single Instruction Multiple Data (SIMD), Multiple Instruction Single Data (MISD) and Multiple Instruction Multiple Data (MIMD) machines respectively. Among these SIMD and MIMD machines are generally used for handling AI problems [21]. In the next section, we briefly outline the features of SIMD and MIMD machines.

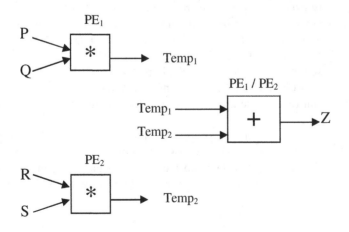

Fig. 1.5: The concept of parallel processing

1.4.1 SIMD and MIMD Machines

An SIMD machine comprises of a single control unit (CU) and a number of synchronous processing elements (PE). Typically there exist two configurations of SIMD architecture. The first configuration employs local (private) memory for each PEs, whereas the second configuration allows flexibility in the selection of memory for a PE by using an alignment network. The user program for both the configurations is saved in a separate memory assigned to the control unit. The CU thus fetches operation codes, decodes and executes the scalar instructions stored in its memory. The decoded vector instructions, however, are mapped to the appropriate PEs by the switching mechanism through a network. Two typical SIMD configurations are presented vide Fig. 1.6 to demonstrate their structural differences.

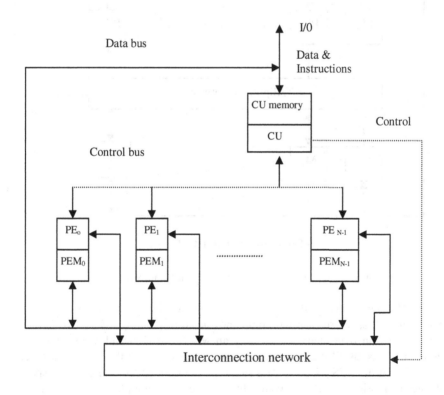

(a) Configuration I (Illiac IV)

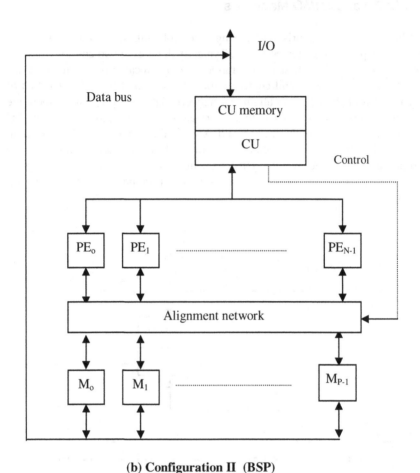

(b) Configuration II (BSP)

Fig. 1.6: Architectural configurations of SIMD array processors

An MIMD machine (as shown in Fig. 1.7), on the other hand employs a number of CUs and PEs, where each CU commands its corresponding PE for executing a specific task on data elements. Usually an MIMD machine allows interactions among the PEs, as all the memory streams are derived from the same data space shared by all the PES. Had the data streams been derived from disjoint subspace of the memories, then we would have called it multiple SISD operation. An MIMD machine is referred to as tightly coupled if the degree of interaction among the PEs is very high. Otherwise they are usually called loosely coupled MIMD machines. Unfortunately, most commercial MIMD machines are loosely coupled.

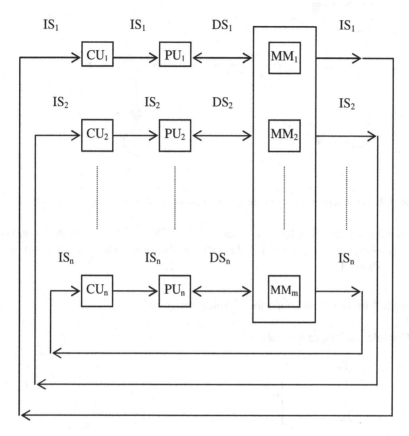

CU: Control unit
PU: Processor unit
MM: Memory module
IS: Instruction Stream
DS: Data Stream

Fig. 1.7: MIMD Computer

1.4.2 Data Flow Architecture

It has already been discussed that the conventional Von Neumann machines fetch instructions from the memory and decodes and executes them in sequence. Because of the sequential organization of the stored programs in memory, possible parallelism among instructions cannot be represented by the program. Dataflow

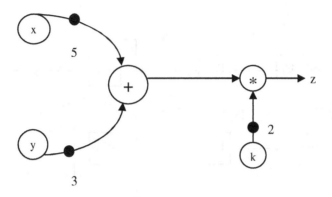

Fig. 1.8: A dataflow graph for the program : z ←(x+y) * k with x = 5, y = 3 and k = 2

architecture, on the other hand represents the possible parallelism in the program by a dataflow graph. Figure 1.8 describes a dataflow graph to represent the program segment:

Example 1.6: Dataflow Graph for a Typical Program.

Consider the program as follows:

$$
\left.
\begin{array}{l}
x := \quad 5 \,; \\
y := \quad 3 \,; \\
k := \quad 2 \,; \\
z := \quad (x + y) * k
\end{array}
\right\}
\tag{1.23}
$$

where := denotes the assignment operator.

Variables in the dataflow graph are usually denoted by circles (containing variable names). The dark dots over the arcs denote the token value of the variables located at the beginning of the corresponding arcs. The operators are mapped at the processing elements depending on their freedom of accessibility. Generally, each processing element has a definite address. The communication of message from one processing element to another is realized by a packet transfer mechanism. Each packet includes the destination address, the input and output parameters and the operation to be executed by the processing elements. A typical packet structure is presented (vide Fig. 1.9) for convenience.

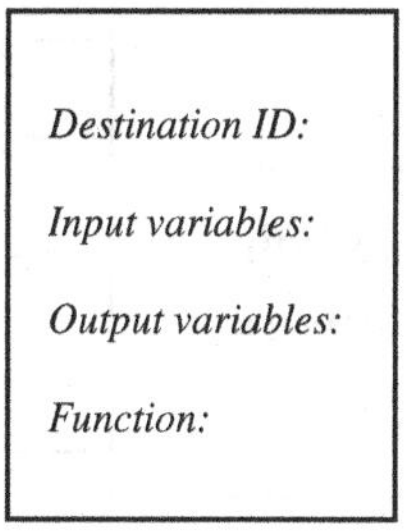

Destination ID:
Input variables:
Output variables:
Function:

Fig. 1.9: A typical packet used for message passing among the PEs

0
0
PE
1
1
N × N
network
PE
(N-1)
(N-1)
PE

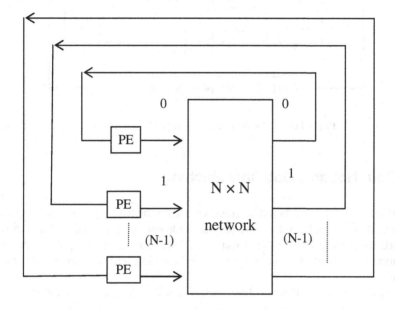

Fig. 1.10: The Arvind dataflow machine

Among the well-known data flow machines, Arvind machine of the MIT and the Manchester University machine are most popular. The basic difference between the two machine architectures lies in the *arbitration unit*. In the Manchester machine, token queues (TQ) are used to streamline tokens from the queues through matching unit (MU), node storage (NS) and the processing units (PU). Transfer of resulting tokens to another PU is accomplished by an exchange switching network. The Arvind machine, however, allows packet transfer through an N x N switching network. The details of the architecture of the two machines are presented in the Fig. 1.10 and Fig. 1. 11 for convenience.

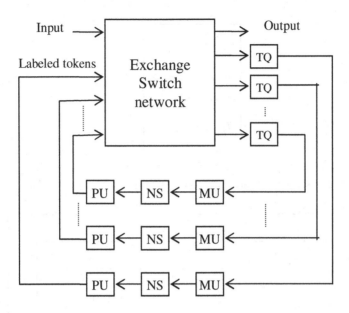

Fig. 1.11: The Manchester machine with multiple ring architecture

1.5 Petri Net as a Dataflow Machine

A Petri net is a directed bipartite graph consisting of two types of nodes: places and transitions. Usually, tokens are placed inside one or more places of a Petri net denoted by circles. The flow of tokens from the input to the output places of a transition is determined by a constraint, called *enabling* and *firing* condition of the transition.

The token-flow in a Petri net has much similarity with the data/token flow in a dataflow architecture. Since token flow in a dataflow machine depends on the presence of the operands (tokens) at a given processing element, token flow may not be continuous in a dataflow machine. Consequently, dataflow architecture is usually categorized under the framework of asynchronus systems. In a Petri net model, the enabling and firing conditions of all the transitions in tandem may not always be satisfied because of resource (token) constraints. This results in asynchronus firing of transitions. Consequently Petri nets too are classified under the framework of parallel asynchronus machines.

The principles of dataflow and asynchronism characteristic of a Petri net being similar to that of a dataflow architecture, Petri nets may be regarded as a special type dataflow machine.

The book attempts to utilize the dataflow characteristics of a Petri net model for realizing the AND-, OR- and Stream-parallelism of a logic program. The scope of Petri nets to model the above types of parallelisms are discussed in detail later in this chapter. The Unification parallelism in logic program does not require any special characteristic of a Petri net model for its realization and fortunately its realization on a Petri net does not invite any additional problem.

1.5.1 Petri Nets — A Brief Review

Petri nets are directed bipartite graphs consisting of two types of nodes called places and transitions. Directed arcs (arrows) connect the places and transitions, with some arcs directed from the places to the transitions and the remaining arcs directed from the transitions to the places. An arc directed from a place p_i to a transition tr_j defines the place to be an input of the transition. On the other hand, an arc directed from a transition tr_k to place p_l indicates it to be an output place of tr_k. Arcs are usually labeled with weights (positive integers), where a k-weighted arc can be interpreted as the set of k-parallel arcs. A marking (state) assigns to each place a non-negative integer. If a marking assigns to place p_i an integer k (denoted by k-dots at the place), we say that p_i is marked with k tokens. A marking is denoted by a vector M, the p_i-th component of which, denoted by $M(p_i)$, is the number of tokens at place p_i. Formally, a Petri net is a 5-tuple, given by

$PN = (P, Tr, A, W, M_0)$
where
$\quad\quad\quad\quad\quad P = \{p_1, p_2, ..., p_m\}$ is a finite set of places,
$\quad\quad\quad\quad\quad Tr = \{tr_1, tr_2, ..., tr_n\}$ is a finite set of transitions,
$\quad\quad\quad\quad\quad A \subseteq (P \times Tr) \cup (Tr \times P)$ is a set of arcs,
$\quad\quad\quad\quad\quad W: A \to \{1, 2, 3, ...\}$ is a weight function,
$\quad\quad\quad\quad\quad M_0: P \to \{0, 1, 2, 3, ...\}$ is the initial marking,
$\quad\quad\quad\quad\quad P \cap Tr = \varnothing$ and $P \cup Tr \neq \varnothing$.

Dynamic behaviour of many systems can be described as transition of system states. In order to simulate the dynamic behaviour of a system, a state or marking in a Petri net is changed according to the following *transition firing* rules:

1) A transition tr_j is enabled if each input place p_k of the transition is marked with at least $w(p_k, tr_j)$ tokens, where $w(p_k, tr_j)$ denotes the weight of the arc from p_k to tr_j.

2) An enabled transition fires if the event described by the transition and its input/ output places actually takes place.

3) A firing of an enabled transition tr_j removes $w(p_k, tr_j)$ tokens from each input place p_k of tr_j, and adds $w(tr_j, p_l)$ tokens to each output place p_l of tr_j, where $w(tr_j, p_l)$ is the weight of the arc from tr_j to p_l.

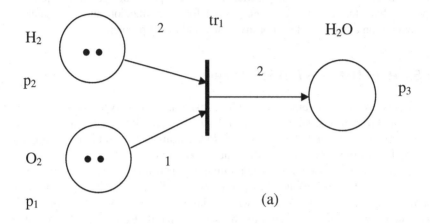

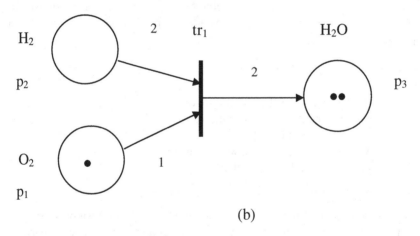

Fig. 1.12: Illustration of transition firing rule in a Petri net. The markings: (a) before the transition firing and (b) after the transition firing

Example 1.7: Consider the well-known chemical reaction: $2H_2 + O_2 = 2H_2O$. We represent the above equation by a small Petri net (Fig. 1.12). Suppose two molecules of H_2 and O_2 are available. We assign two tokens to the places p_2 and p_1 representing H_2 and O_2 molecules respectively. The place p_3 representing H_2O is initially empty (Fig. (1.12(a)). Weights of the arcs have been selected from the given chemical equation. Let the tokens residing at place H_2 and O_2 be denoted by $M(p_2)$ and $M(p_1)$ respectively. Then we note that

$$M(p_2) = W(p_2, tr_1) \text{ and } M(p_1) > W(p_1, tr_1).$$

Consequently, the transition tr_1 is enabled, and it fires by removing two tokens from the place p_2 and one token from place p_1. Since the weight $W(tr_1, p_3)$ is 2, two molecules of water will be produced, and thus after firing of the transition, the place p_3 contains two tokens. Further, after firing of the transition tr_1, two molecules of H_2 and one molecule of O_2 have been consumed and only one molecule of O_2 remains in place p_1.

The dynamic behaviour of Petri nets is usually analyzed by a state equation, where the tokens at all the places after firing of one or mores transitions can be visualized by the marking vector M. Given a Petri net consisting of n transitions and m place.

Let

A = $[a_{ij}]$ be an (n × m) matrix of integers, called the *incidence matrix*, with entries

$$a_{ij} = a_{ij}^+ - a_{ij}^-$$

where

$a_{ij}^+ = w(tr_i, p_j)$ is the weight of the arc from transition tr_i to place p_j,

and $a_{ij}^- = w(p_j, tr_i)$ is the weight of the arc to transition tr_i from its input place p_j.

It is clear from the transition firing rule described above that a_{ij}^-, a_{ij}^+ and a_{ij} respectively represent the number of tokens removed, added and changed in place j when transition tr_i fires once. Let M be a marking vector, whose j-th element denotes the number of tokens at place p_j. The transition tr_i is then enabled at marking M if

$$a_{ij}^- \leq M(j), \text{ for } j = 1, 2, ..., m. \tag{1.24}$$

In writing matrix equations, we write a marking M_k as an (m × 1) vector, the j-th entry of which denotes the number of tokens in place j immediately after the k-th firing in some firing sequence. Let u_k be a control vector of (n × 1) dimension consisting of (n−1) zeroes and a single 1 at the i-th position, indicating that transition tr_i fires at the k-th firing. Since the i-th row of the incidence matrix A represents the change of the marking as the result of firing transition tr_i, we can write the following state equation for a Petri net:

$$M_k = M_{k-1} + A^T u_k, \ k = 1,2,... \tag{1.25}$$

Suppose we need to reach a destination marking M_d from M_0 through a firing sequence $\{u_1, u_2, ..., u_d\}$. Iterating k = 0 to d in incremental steps of 1, we can then write:

$$\left.\begin{array}{l} M_1 = M_0 + A^T u_1 \\ M_2 = M_1 + A^T u_2 \\ \dots \ \dots \ \dots \ \dots \\ \dots \ \dots \ \dots \ \dots \\ M_{d-1} = M_{d-2} + A^T u_{d-1} \\ M_d \ = M_{d-1} + A^T u_d \end{array}\right\} \tag{1.26}$$

Equating the left hand sum with the right hand sum of the above equations we have:

$$M_d = M_0 + \sum_{k=1}^{d} A^T u_k \tag{1.27}$$

$$\text{or, } M_d - M_0 = A^T \sum_{k=1}^{d} u_k \tag{1.28}$$

$$\text{or, } \Delta M = A^T x, \tag{1.29}$$

where

$$\Delta M = M_k - M_0, \tag{1.30}$$

and

$$x = \sum_{k=1}^{d} u_k. \tag{1.31}$$

Here x is a $(n \times 1)$ column vector of non-negative integers, and is called the *firing count vector* [24]. The i-th entry of x denotes the number of times that transition tr_i must fire to transform M_0 to M_d.

Example 1.8: The state equation (1.25) is illustrated with the help of Fig. 1.13. It is clear from the figure that $M_0 = [\, 2 \ 0 \ 1 \ 0 \,]^T$. After firing of transition tr_3, we obtain the resulting marking M_1 by using the state equation as follows:

$$M_1 = M_0 + A^T u_1$$

$$= [\, 2 \ 0 \ 1 \ 0 \,]^T + \begin{pmatrix} -2 & 1 & 1 \\ 1 & -1 & 0 \\ 1 & 0 & -1 \\ 0 & -2 & 2 \end{pmatrix} [\, 0 \ 0 \ 1 \,]^T$$

$$= [\, 3 \ 0 \ 0 \ 2 \,]^T.$$

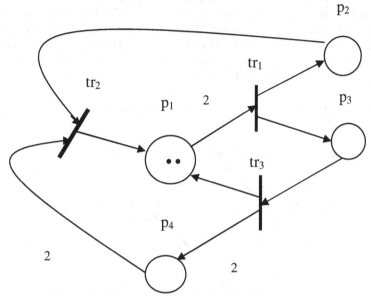

Fig. 1.13: A Petri net used to illustrate the state equation

1.6 Parallelism in Logic Programs — A Review

A typical logic program (vide section 1.3.1) is a collection of Horn clauses. The resolution process presented in section 1.2 was illustrated with *Select Linear Definite (SLD)* clauses. Under this scheme, a given set of clauses S including the query is the input to a resolution system, where two clauses having oppositely signed common literals, present in the body of one clause Cl_1 and the head of another clause Cl_2, are resolved to generate a resolvent Cl_3 . The Cl_3 is then resolved with another clause from S having oppositely signed common literals. The process terminates when no further resolution is feasible or a null clause is produced yielding a solution for the argument terms of the predicate literals. The whole process is usually represented by a tree structure, well-known as *SLD-tree*. The SLD-tree thus allows binary resolution of clauses, with the resolvent carried forward for resolution with a third clause.

The SLD-resolution is a systematic tool for reasoning in a logic program realized on a uniprocessor architecture. However, the principle can easily be extended for concurrent resolution of multiple program clauses. Various alternative formulations for concurrent resolution of multiple program clauses are available in the literature [22]. One typical scheme is briefly outlined. In this scheme, we first select m number of pair of clauses (including the goal) that can participate in the resolution process. If such m pairs are available, then we would have (m/2) number of resolvents. If the resolvents can again be paired so that they are resolvable we could find (m/4) resolvents, and so on, until we find two clauses which on resolution finally may give rise to a null clause. Had it been so, we require only k units of time, to resolve m = 2^k clauses. The total time of resolution k = $\log_2$ (m) can be reduced further if we can resolve more than two clauses together by some mechanism. In fact this too can be realized, if appropriate hardware/software resources are available. The above two types of concurrent resolution is illustrated in example 1.9:

Example 1.9: The SLD-tree for the given logic program:

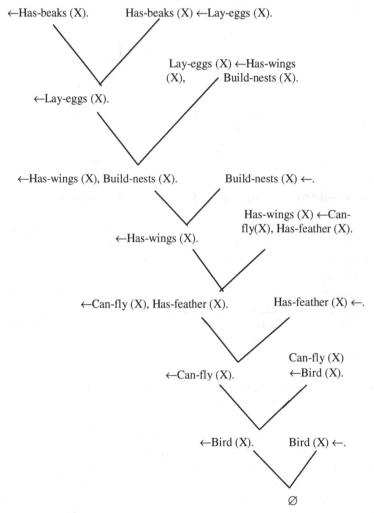

Fig. 1.14: An SLD resolution-tree depicting resolution of two clauses, and the resolvent been passed on for resolution with a third clause until a null clause is derived, or no further resolution is possible

Logic Program:

$$\text{Can-fly (X)} \leftarrow \text{Bird (X).} \tag{1.32}$$
$$\text{Bird (X)} \leftarrow. \tag{1.33}$$
$$\text{Has-feather (X)} \leftarrow. \tag{1.34}$$
$$\text{Has-wings (X)} \leftarrow \text{Can-fly (X) , Has-feather (X).} \tag{1.35}$$
$$\text{Has-beaks (X)} \leftarrow \text{Lay-eggs (X).} \tag{1.36}$$
$$\text{Lay-eggs (X)} \leftarrow \text{Has-wings (X), Build-nests (X).} \tag{1.37}$$
$$\text{Build-nests (X)} \leftarrow. \tag{1.38}$$

$$\text{Query: } \leftarrow \text{Has-beaks (X).} \tag{1.39}$$

For the given logic program, the SLD resolution tree is given in Fig. 1.14.

When we take two clauses concurrently as available for resolution, the resolution tree looks like Fig. 1.15 as illustrated here.

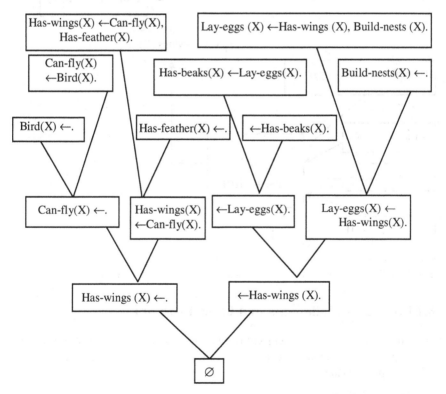

Fig. 1.15: Illustrating the process of resolving multiple (>2) program clauses together

The process of composite resolution of multiple program clauses is illustrated vide Fig. 1.16.

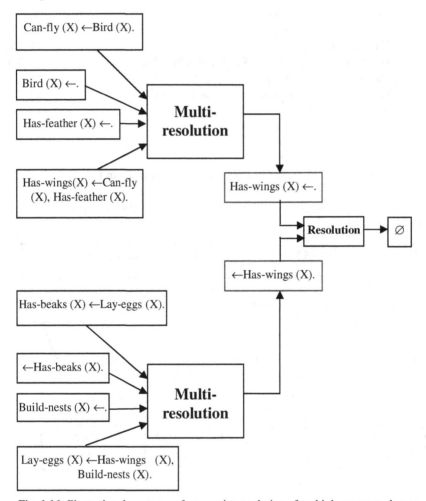

Fig. 1.16: Illustrating the process of composite resolution of multiple program clauses

1.6.1 Possible Parallelisms in a Logic Program

Besides the above forms of concurrent resolution of clauses, there exist four other types of parallelisms in a logic program. These are AND-, OR-, Stream- and Unification-parallelisms.

(a) AND–parallelism

The literals (predicates) separated by commas in the body of a Horn clause are usually called *AND-literals*. The AND-literals of a clause may be searched against the heads of the available clauses for resolution. The concurrent resolution of the AND-literals of a clause with the heads of other clauses is usually called *AND-parallelism*. Example 1.10 illustrates AND-parallelism in a typical logic program.

Example 1.10: Let us consider the following logic program.

$$\text{Parent (M, F, X)} \leftarrow \text{Father (F, X), Mother (M, X).} \qquad (1.40)$$
$$\text{Mother (jaya, tom)} \leftarrow. \qquad (1.41)$$
$$\text{Mother (ipsa, bil)} \leftarrow. \qquad (1.42)$$
$$\text{Father (asit, tom)} \leftarrow. \qquad (1.43)$$
$$\text{Father (amit, bil)} \leftarrow. \qquad (1.44)$$

$$\text{Query: } \leftarrow\text{Parent (M, F, bil).} \qquad (1.45)$$

In Fig. 1.17 the clause '←Father (F, bil), Mother (M, bil).' is instantiable by clauses (1.42) and (1.44) concurrently. Thus a concurrent resolution of three clauses take place jointly resulting in a null clause. The results of instantiation of the variables F and M in the present case are F = amit and M = ipsa.

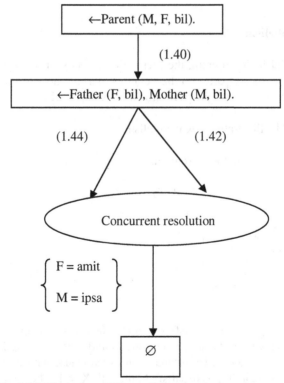

Fig. 1.17: Illustrating AND-parallelism in a logic program

In case the body of the clause contains AND-literals having shared variables, then a binding conflict in the process of instantiation might arise. For instance, consider the following logic program

$$F (X) \leftarrow A (X), B (X). \tag{1.46}$$
$$A (1) \leftarrow. \tag{1.47}$$
$$B (2) \leftarrow. \tag{1.48}$$

Here, the possible instantiation of the AND literals are $X = 1$ or $X = 2$. But both $X = 1$ and $X = 2$ jointly cannot hold as it violates (1.46).

In AND-parallelism, where the goals/sub-goals that include shared variables, are allowed to be resolved independently in parallel, is called *unrestricted AND-parallelism*. A considerable runtime overhead is incurred for synchronization of shared variables and filtering the set of variable bindings to answer the query in such systems. To avoid such overhead, AND-parallelism is allowed when variable bindings are conflict-free. Such AND-parallelism is referred to as the *restricted AND-parallelism*. For the implementation of conflict freedom, a program annotation is necessary to denote which goals/sub-goals produce or consume variables.

(b) OR-parallelism

In a sequential logic program, the literals in the body of a clause are unified in order with the head part of other clauses during the process of resolution. Consider, for instance, the program shown in example 1.11.

Example 1.11: Illustrating OR-parallelism.

Consider the following logic program.

$$1. \quad \text{Main} \leftarrow A (X), P (X). \tag{1.49}$$
$$2. \quad A (1) \leftarrow b, c, d. \tag{1.50}$$
$$3. \quad A (2) \leftarrow e, f, g. \tag{1.51}$$
$$4. \quad A (3) \leftarrow h, i. \tag{1.52}$$
$$5. \quad P (3) \leftarrow p, c. \tag{1.53}$$

In the above program to satisfy the goal: Main, one attempts to unify the first sub-goal A (X) with A (1), and then he/she should start searching P (1) in the head of the subsequent clauses. Unfortunately no such clause with P (1) in the head is available; so the search fails to satisfy Main with $X = 1$. The same process is then repeated for $X = 2$, but main is not satisfied again as P (2) is not available as the head of some clauses. The goal, however, succeeds by unifying A (X) and P (X) with heads A (3) and P (3) respectively.

However, given sufficient computing resources, it is possible to perform the unification of A (X) with A (1), A (2) and A (3) in parallel. Such concurrent unification of A (X) with OR-clauses A (1), A (2) and A (3) in parallel is called *OR-parallelism*. The difficulty of OR-parallelism, with respect to the last example, is the propagation of the correct bindings of variable X to P (X). This, however, calls for some knowledge about the existence of P (3) as a head of some clauses. Perhaps, by allowing concurrency of AND- as well as OR-parallelism, this could be made possible in future logic programming machines.

(c) Stream-parallelism

Stream-parallelism occurs in a logic program, when the literals pass a stream of variable bindings to other literals, each of which is operated on concurrently. Literals producing the variable bindings are called *producers*, while the literals that use the bound value of variables are called *consumers* (vide [12]). Example 1.12 illustrates Stream-parallelism.

Example 1.12: Illustrating Stream-parallelism.

1.	Main ←Int (N), Test (N), Print (N).	(1.54)
2.	Int (0) ←.	(1.55)
3.	Int (N) ←(M), N is M + 1.	(1.56)

In the last program to satisfy Main, one needs to satisfy Int (N), Test (N) and Print (N) in sequence. Once Int (N) is unified with Int (0) ←. , the value of the parameter N = 0 is passed on to Test (N) and then Print (N) in succession.

Thus when Test (N) and Print (N) are executed with old bindings of N, new bindings of N may be generated concurrently by unifying Int (N) of clause (1.54) with the head of the clause (1.56). Such parallelism where the clauses in the body of a clause are unified with the result of binding, of their preceding clauses in the body is called S*tream-parallelism*. Stream-parallelism has similarity with pipelining. Test (N) and Print (N), for instance, are similar with processes, where they wait for the data streams to be produced by the preceding process Int (N).

(d) Unification-parallelism

In unification-parallelism, the terms in the argument of a predicate are instantiated in parallel with the corresponding terms of another predicate. For instance, the Petri net corresponding to the logic program presented allows binding of the variables X and Y in the arc function concurrently with the tokens a and b respectively at place P.

Consider the following logic program.

Logic Program:

$$1.\ R\,(Z, X) \leftarrow P\,(X, Y), Q\,(Y, Z).$$
$$2.\ P\,(a, b) \leftarrow.$$
$$\left.\right\} \qquad\qquad (1.57)$$

Here, the two clauses can resolve if the predicate P in both the clauses can be unified. In case of unification-parallelism, the variables (X and Y) in the argument of P under the first clause are bound concurrently with the constants (a and b) in the argument of P under the second clause.

1.7 Scope of Parallelism in Logic Programs using Petri Nets

After a careful observation of various types of logic programs, we arrive at the conclusion that Petri net model truly resemble all the features that a concurrent logic program requires for execution. The classical model of Petri net, we introduced in section 1.5, however, needs an extension for its suitability to realize the parallelisms in a logic program.

A number of authors have already suggested several models of Petri net for parallel realization of logic programs. A detailed discussion on this is given in section 2.8. For the sake of completeness of this chapter, we have briefly outline one typical model of Petri net and demonstrate its application in synthesis of AND-, OR-, Stream-parallelism of logic programs.

Example 1.13: This example illustrates a logic program using Petri net

Logic Program:

$$R(Z, X), S(X, Z) \leftarrow P(X, Y), Q(Y, Z).$$
$$P(a, b) \leftarrow.$$
$$Q(b, c) \leftarrow.$$
$$\neg R(c, a) \leftarrow.$$

The Petri net representation of the above logic program is given by the Fig. 1.18.

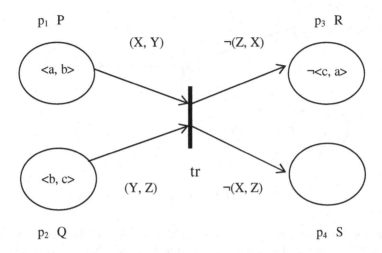

Fig. 1.18: The Petri net for a given logic program

AND-parallelism using Petri net

The literals present in the body part of a clause are referred to as AND-literals. During resolution the AND-literals of a given clause may be searched against the literals present in the heads of other available clauses.

In AND-parallelism the binding of terms AND-literals takes place concurrently with those of the literals present in the heads of other clauses. The example below illustrates the concept of AND-parallelism.

Consider the following three program clauses.

$$F(X) \leftarrow A(X), B(Y). \tag{1.58}$$
$$A(1) \leftarrow. \tag{1.59}$$
$$B(1) \leftarrow. \tag{1.60}$$

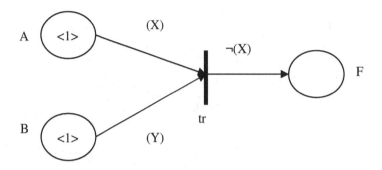

Fig. 1.19: A Petri net illustrating AND-parallelism

For simplicity we consider only one literal in the heads of the second and third clauses. It is clear that the variable X shared by the AND-literals may be searched with the literals present in the heads of the second and the third clauses. It is indeed important to note that an extended Petri net supports AND-parallelism as the input arc functions of a transition can be matched concurrently with the available resources of its corresponding input places.

After the mapping of the three clauses onto the Petri net is over, the network looks like Fig. 1.19. Hence the input arc functions X and Y of the transition tr can be matched with the tokens residing at place A and B concurrently. It may be noted that the tokens appeared at the places A and B of the Petri net because of mapping of the rules (1.59) and (1.60) onto the Petri net respectively. Thus it is clear that, in absence of conflict in binding, AND-parallelism can easily take place in a Petri net.

When the AND-literals of a clause contain shared variables, a binding conflict in the process of instantiation may arise. For instance, consider the following program containing three program clauses.

$$F(X) \leftarrow A(X), B(X). \tag{1.61}$$
$$A(1) \leftarrow . \tag{1.62}$$
$$B(1) \leftarrow . \tag{1.63}$$

When the variable X of literals A and B corresponding to the first clause is attempted to match against A(1) and B(1) of the second and third clauses, the resolution succeeds and there is no restriction in AND- parallelism when AND-clauses share common variables.

OR-Parallelism using Petri net

In case of OR-parallelism a literal present in the body of one clause may be searched concurrently against the literals present in the heads of more than one clause. The latter clauses are usually referred to as OR-clauses. For example consider the following program.

$$F(X) \leftarrow A(X), B(Y). \qquad (1.64)$$
$$A(1) \leftarrow. \qquad (1.65)$$
$$A(2) \leftarrow. \qquad (1.66)$$
$$B(1) \leftarrow. \qquad (1.67)$$

Here, the variable X present in the body of the first clause is matched concurrently with the arguments of the literal A in the second and third clauses. It may be mentioned here that the Petri net representation of the above program clauses, (vide Fig. 1.20) allows concurrent matching of the arc function variable X with the tokens <1>, <2> residing at place A. However, such concurrent matching requires additional system resources.

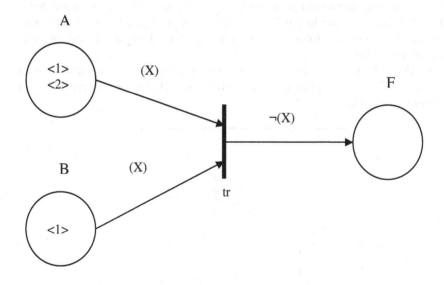

Fig. 1.20: A Petri net illustrating OR-parallelism

Stream-Parallelism using Petri net

In SLD resolution, the resolvent of two clauses participates in a subsequent resolution with a third clause. In case the second resolution takes place on a sequential stream of tokens generated by the first resolution we say that a pipeline exists between the two successive resolutions.

Let p_i for i = 1, 2, ...,m be the input places of a transition tr_k and p_j for j = m+1, m+2,....., n be the output places of the same transition. Also assume that there exists another transition tr_{k+1}, which has an input place p_j for some j, where m+1 < j < n. Under this case we call the two transitions to be in pipeline. In case the transition tr_{k+1} waits for a number of firing of transition tr_k to produce a token, we say that a Stream-parallelism persists in the Petri net model.

The example below illustrates the realization of Stream-parallelism in the Petri net model.

$$Int(0) \leftarrow. \tag{1.68}$$
$$Int(M+1) \leftarrow Int(M). \tag{1.69}$$
$$Compute(N \times N) \leftarrow Int(N). \tag{1.70}$$

The above logic program generates integers N and computes its square. Since these two processes are in pipeline, and integer N is generated for all successive values of N counting from zero, the corresponding Petri net model includes Stream-parallelism.

In Fig. 1.21 the place-transition pair p_1-tr_1 corresponds to the generation of integers: 1, 2, ..., ∞ at place p_1. The resulting token at place p_1 is then used up for computing its square.

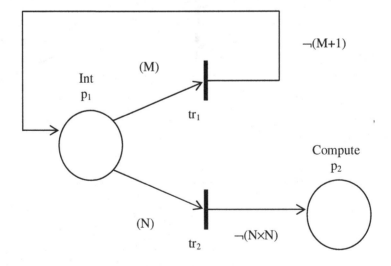

Fig. 1.21: A Petri net illustrating Stream-parallelism

Unification-Parallelism using Petri net

In a Petri net model the constant arguments of a predicate are usually mapped as tokens of places. Thus variable arguments of the same predicate used in another rule is represented as arc functions of an arc connected between the place representing the predicates and the transition describing the rule. In the process of resolution of a clause with a fact of constant arguments, the variables present in the literal of the first clause is unified with the same literal present in the fact. In the Petri net model the variables in the arc function are matched with the tokens of the connected places. When an arc function contains a number of variables the instantiation of each variable takes place with the position-wise constant of the tokens.

The Petri net in Fig. 1.22 represents the logic program given below.

Logic program:

$$R(Y, X) \leftarrow P(X, Y).$$ (1.71)
$$P(a, b) \leftarrow.$$ (1.72)

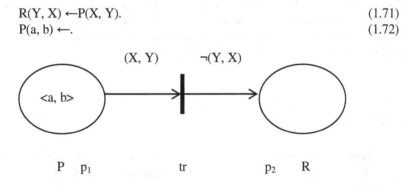

Fig. 1.22: Petri net demonstrating Unification-parallelism

In Fig. 1.22, the predicate P(X, Y) of (1.71) is unified with predicate P(a, b) of (1.72). The instantiation of all the variables can be done in the Petri net model concurrently with sufficient system resources. The variable argument of predicate P of (1.71) is represented by an arc function (X, Y), and the constant token (a, b) of (1.72) is denoted by a token of place p_1. The instantiation of variables X and Y can be done concurrently with the tokens a and b located at place p_1.

In this book we consider Petri net models capable of representing multiple antecedent and multiple consequent clauses. Usually the commas present in the antecedent clauses denote conjunction and in consequent clauses denote disjunction. Thus in presence of tokens at all but one input-output places of an enabled transition, the transition will fire generating a new token. Such firing of transition includes typical AND- and a different type of OR-parallelism. Here, independent facts mapped at the output places of the transition behave like typical OR-clauses, and a set of concurrent resolution takes place between the OR-clauses and a given rule containing those literals present in the OR-clauses as consequents.

Unification-parallelism can always be maintained in the Petri net model, and Stream-parallelism exists only when the network includes pipelined transitions where a transition in the pipeline waits for the other to generate a sequence of tokens.

1.8 Conclusions

The chapter explores different models of parallel architecture and finally identifies Petri nets as a suitable architecture for concurrent resolution of logic programs. The execution of a logic program using resolution principles introduced and the scope of different parallelisms in a logic program are identified. Finally the chapter comes to an end with a discussion on the realization of possible parallelisms in a logic program using Petri nets.

Exercises

1. Given two clauses $P \vee Q_1$ and $\neg Q_2 \vee R$ with $P = A (X, Y)$, $Q_1 = B (Y, X)$, $Q_2 = B (b, a)$, $R = C (Y, X)$. Determine the substitution set S and evaluate $(P \vee R)[S]$.

    [**Hints:** From the resolution principle, we can see that after unifying Q_1 and Q_2, the resolvent $(P \vee R)[S] = A (a, b) \vee C (b, a)$.
    Here, the substitution set $S = \{a/X, b/Y\}$.]

2. Unify the following two predicates:

    i) $P_1 = $ Loves $(X,$ son-of $(X))$ and
       $P_2 = $ Loves $($mam, $Y)$.

    ii) $P_1 = P(a, X, f(g(X)))$ and
        $P_2 = P(Z, f(Z), f(U))$.

[**Hints:**

i) Here, these two predicates can be unified as follows:
$P = P_1 = P_2 = $ Loves (mam, son-of (mam))
where the substitution is given by
$S = \{$mam/X, son-of(mam)/Y$\}$.

ii) Here, these two predicates can be unified as follows:
$P = P_1 = P_2 = $ P(a, f(a), g(f(a))
where the substitution is given by
$S = \{$a/Z, f(a)/X, g(f(a))/U$\}$.]

3.   Given the logic program and the query. How will you employ resolution principle to answer the query?

Logic program:

Triangle (XYZ) ←Has-three-sides (XY, YZ, ZX), Has-three-angles (∠XYZ, ∠YZX, ∠ZXY), Equal (∠XYZ + ∠YZX + ∠ZXY, 180°).
Has-three-sides (ab, bc, ca) ←.
Has-three-angles (∠abc, ∠bca, ∠cab) ←.
Equal (∠abc + ∠bca + ∠cab, 180°) ←.
Query: ←Triangle (abc).

[**Hints:** Let P: Has-three-sides
                Q: Has-three-angles
                R: Equal
                T: Triangle

      ∴ The logic program:

T (XYZ) ←P (XY, YZ, ZX), Q (∠XYZ, ∠YZX, ∠ZXY), R (∠XYZ + ∠YZX + ∠ZXY, 180°).

≡ ¬(P (XY, YZ, ZX), Q (∠XYZ, ∠YZX, ∠ZXY), R (∠XYZ + ∠YZX + ∠ZXY)) ∨ T (XYZ)

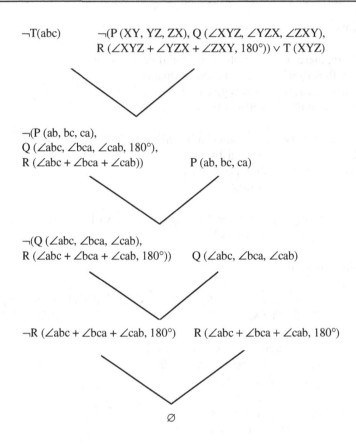

¬T(abc)   ¬(P (XY, YZ, ZX), Q (∠XYZ, ∠YZX, ∠ZXY),
                R (∠XYZ + ∠YZX + ∠ZXY, 180°)) ∨ T (XYZ)

¬(P (ab, bc, ca),
Q (∠abc, ∠bca, ∠cab, 180°),
R (∠abc + ∠bca + ∠cab))   P (ab, bc, ca)

¬(Q (∠abc, ∠bca, ∠cab),
R (∠abc + ∠bca + ∠cab, 180°))   Q (∠abc, ∠bca, ∠cab)

¬R (∠abc + ∠bca + ∠cab, 180°)   R (∠abc + ∠bca + ∠cab, 180°)

∅

Fig. 1.23: The resolution tree of the given logic program

P (ab, bc, ca) ←.
≡ P (ab, bc, ca)

Q (∠abc, ∠bca, ∠cab) ←.
≡ Q (∠abc, ∠bca, ∠cab)

R (∠abc + ∠bca + ∠cab, 180°) ←.
≡ R (∠abc + ∠bca + ∠cab, 180°)

and the query: ←T (abc).
≡ ¬T (abc)

Fig. 1.23 represents the resolution tree for the given logic program.

∴ According to the resolution principle, resolution of

¬(P (XY, YZ, ZX), Q (∠XYZ, ∠YZX, ∠ZXY), R (∠XYZ + ∠YZX + ∠ZXY, 180°)) ∨ T (XYZ)

and

¬T (abc)

yields a resolvent

¬(P (ab, bc, ca), Q (∠abc, ∠bca, ∠cab), R (∠abc + ∠bca + ∠cab, 180°)).

 The resolvent then resolves with

P (ab, bc, ca), Q (∠abc, ∠bca, ∠cab)

and R (∠abc + ∠bca + ∠cab, 180°)

successively to yield a null result which signifies that

'abc is a triangle' is the answer to the query.]

4.   Given the following logic program and the query. Show that the contents of the stack corresponding to each node in the process of expanding the resolution tree.

Logic program:

      1. R (Z, X) ←P (X, Y), Q (Y, Z).
      2. P (a, c) ←.
      3. P (a, b) ←.
      4. Q (b, c) ←.

      Query: ←R (c, a).

[Hints:

Based on the principles outlined in section 1.3.2, the resolution tree is constructed using stack (Fig. 1.24).

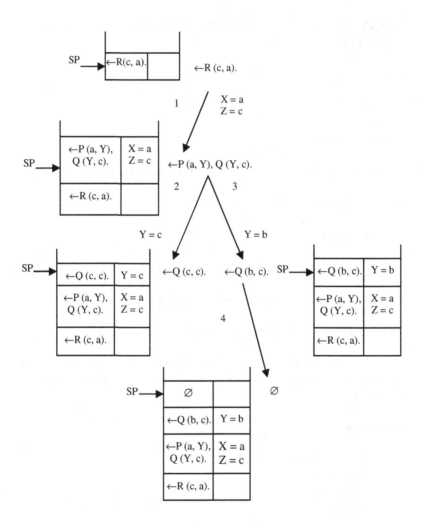

Fig. 1.24: The resolution tree showing the contents of the stack corresponding to each node]

5.   Draw the SLD-tree for the following logic program and mark on the tree the part of the search space that remains unexplored because of using the CUT(!) statement.

Logic program:

Cl_1: P ←Q, R.

....

....

Cl_4: Q ←S, !, T.

....

....

Cl_7: S ←.

and

Goal: ←P.

[**Hints:** Fig. 1.25 shows to control backtracking by using CUT.

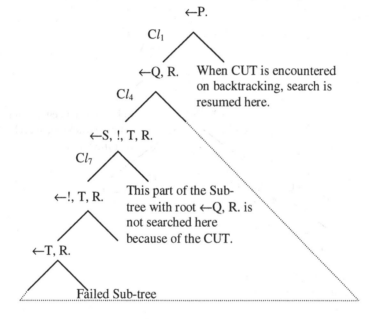

Fig. 1.25: Controlling backtracking by using CUT

Literals preceding CUT are unifiable with the same literals in the head of other clauses. So, ! is automatically satisfied. Since ←T, R. cannot be resolved with any more clauses, the control returns to the root of the tree ←P. for generating alternative solution.]

6. Construct the SLD-tree for the following logic program and show that when a failure occurs before the 'CUT' statement the control returns to the parent of the clause under consideration.

Logic Program:

      Cl_1: A ←B, C.

      ....
      ....

      Cl_4: B ←D, E, !, F.

      ....
      ....

      Cl_7: D ←.
      Cl_8: B ←M.
      and
      Goal: ←A.

[Hints:

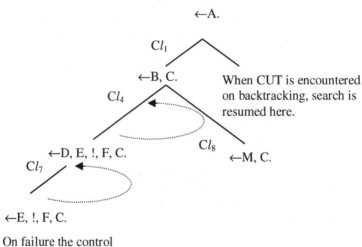

Fig. 1.26: Controlling backtracking by using CUT

On Failure at '←E, !, F, C.' the control returns to its parent and attempts to expand it. Unfortunately the parent, too, cannot be expanded; so the control again starts exploring the grandparent of '←E, !, F, C.' node and fortunately can expand the node '←B, C.' by Cl_8.]

7.   'Fail' is another built-in predicate used in conjunction with 'CUT' in order to intentionally cause a failure of the root clause and force the control to backtrack to the root for finding alternative solutions.

Consider the logic program involving 'CUT' with 'Fail' predicate shown as follows. Show the steps of backtracking on the SLD-tree for the programs and hence find the solution to the problem.

Logic program:

Cl_1: Tax-payer(X) ←Annual-inc(X, Earnings), Earnings ≤ 30000, !, Fail.
Cl_2: Tax-payer(X) ←Annual-inc(Family-members-of(X), Earnings),
                     Earnings < 40000, !, Fail.
Cl_3: Tax-payer(X) ←Annual-inc(X, Earnings), Earnings > 30000.
Cl_4: Annual-inc(titir, 25000) ←.
Cl_5: Annual-inc(Family-members-of(tunir), 30000) ←.
Cl_6: Annual-inc(tapur, 50000) ←.

[**Hints:** The tree for the tax-payer problem is constructed by the following policy.

When a predicate before a 'CUT' predicate is satisfied, we drop the predicate from the list using SLD-resolution. The 'CUT' is automatically satisfied on getting a 'FAIL' after a 'CUT', the control returns to the root of the tree for exploring the possibility of alternative solutions. When the clause at the leading edge of the tree is satisfied, a null clause is generated, causing the termination of the tree.

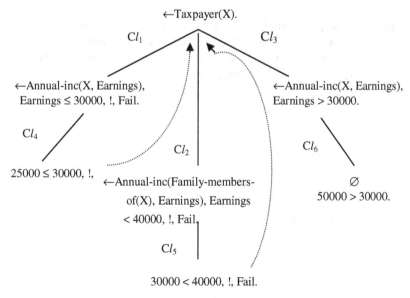

Fig. 1.27: Controlling backtracking using 'CUT' and 'Fail']

8. Identify the stable points for the following logic programs:

a) $\neg p \leftarrow \neg q.$
   $p \leftarrow.$

b) $q \leftarrow p.$
   $r \leftarrow q.$
   $\neg r \leftarrow.$

c) $r \leftarrow (q \leftarrow p).$
   $\neg r \leftarrow.$

d) $p \leftarrow q.$
   $q \leftarrow p.$

[**Hints:**

a) We obtain from the first clause the interpretations $(p, q) \in \{(0, d), (d, 1)\}$ where d denotes *a don't care state* for the respective proposition. The interpretation from the second clause is $\{(1, d)\}$. The intersection of these two interpretations yield a stable point

$$(p, q) = (1, 1).$$

b) The interpretations from the first clause are $(p, q, r) = \{(0, d, d), (d, 1, d)\}$. The interpretations from the second and the third clauses are $\{(d, 0, d), (d, d, 1)\}$ and $\{(d, d, 0)\}$. The intersection of the above three interpretations yield a stable point

$$(p, q, r) = (0, 0, 0).$$

c)

$$r \leftarrow (q \leftarrow p).$$
$$\equiv r \vee \neg (q \leftarrow p).$$
$$\equiv r \vee \neg (q \vee \neg p).$$
$$\equiv r \vee (p \wedge \neg q).$$
$$\equiv (p \vee r) \wedge (\neg q \vee r).$$

Truth Table of the expression is given below:

p	q	r	r ←(q ←p).
0	0	0	0
0	0	1	1
0	1	0	0
0	1	1	1
1	0	0	1
1	0	1	1
1	1	0	0
1	1	1	1

Therefore, the intersection of these two interpretations yield a stable point

$(p, q, r) = (1, 0, 0)$.

d)

p ←q.
The stable point for the above expression
$(p, q) \in \{(0, 0), (1, 0), (1, 1)\}$

q ←p.
The stable point for the above expression
$(p, q) \in \{(0, 0), (0, 1), (1, 1)\}$

Therefore, the stable point (p, q) belongs to the intersection of the above two interpretations, i.e.,
$(p, q) \in \{(0, 0),(1, 1)\}.]$

9. Given a dynamical system of two propositions p (t) and q (t) where t denotes time. Determine the stable points of the dynamics.

$p (t + 1) = 1$, if f (p (t), q (t)) = 1;
            = 0, otherwise.

$q (t + 1) = 1$, if g (p (t), q (t)) = 1;
            = 0, otherwise.

where f (p (t), q (t)) = q (t) ←(q (t) ←p (t)).
and g (p (t), q (t)) = ¬q (t) ←(¬q (t) ←¬p (t)).

[Hints: $f(p(t), q(t)) \equiv q(t) \leftarrow (q(t) \leftarrow p(t))$.

$\equiv q(t) \vee \neg(q(t) \leftarrow p(t))$.

$\equiv q(t) \vee \neg(q(t) \vee \neg p(t))$.

$\equiv q(t) \vee (\neg q(t) \wedge p(t))$.

$\equiv (q(t) \vee \neg q(t)) \wedge (q(t) \vee p(t))$.

$\equiv 1 \wedge (p(t) \vee q(t))$.

$\equiv (p(t) \vee q(t))$.

$g(p(t), q(t)) \equiv \neg q(t) \leftarrow (\neg q(t) \leftarrow \neg p(t))$.

$\equiv \neg q(t) \vee \neg(\neg q(t) \leftarrow \neg p(t))$.

$\equiv \neg q(t) \vee \neg(\neg q(t) \vee \neg(\neg p(t)))$.

$\equiv \neg q(t) \vee (q(t) \wedge \neg p(t))$.

$\equiv (\neg q(t) \vee q(t)) \wedge (\neg q(t) \vee \neg p(t))$.

$\equiv 1 \wedge (\neg q(t) \vee \neg p(t))$.

$\equiv \neg p(t) \vee \neg q(t)$.

Now, $p(t+1) = f(p(t), q(t)) = (p(t) \vee q(t))$.

$q(t+1) = g(p(t), q(t)) = \neg p(t) \vee \neg q(t)$.

Let $p(t) = p^*(t)$ and $q(t) = q^*(t)$ be the stable points, if any.

$p^*(t) = p^*(t) \vee q^*(t)$

$\Rightarrow \quad p^*(t) = p^*(t)$ or $p^*(t) = q^*(t)$.

$q^*(t) = \neg p^*(t) \vee \neg q^*(t)$

$\Rightarrow \quad q^*(t) = \neg p^*(t)$ or $q^*(t) = \neg q^*(t)$ which is absurd.

$p^*(t) = p^*(t) \vee q^*(t) = Max(p^*(t), q^*(t))$ holds

if $p^* = Max(p^*(t), q^*(t)) = p^*(t)$

i.e., $Max(p^*(t), q^*(t)) = p^*(t)$

or, $p^*(t) \geq q^*(t)$

or, $(p^*(t), q^*(t)) \in \{(1, 0), (1, 1)\}$.   (1.73)

Again, $q^*(t) = \neg p^*(t) \vee \neg q^*(t) = Max(\neg p^*(t) \vee \neg q^*(t))$

if $Max(\neg p^*(t) \vee \neg q^*(t)) = \neg p^*(t) = q^*(t)$

Solution of $\neg p^*(t) = q^*(t)$ are $\{(0, 1), (1, 0)\}$.   (1.74)

The common solution of (1.71) and (1.72) is $(p, q) = (1, 0)$.**]**

10. Consider an arithmetic unit inside a CPU, capable of multiplying and adding two hexadecimal numbers. Given a program

$X = a + b$;
$Y = a - b$;
$Z = X * Y$;
Print Z.

Using two processing units of the given type how many computational cycles will you require to execute the program? Is there any time saving if we employ three such processing elements?

[**Hints:**

Using two processing units of the given type, two computational cycles will be needed to execute the program as the two processing units will be used to evaluate the values of X and Y in parallel and then any one of the processor will compute the value of Z.

As the evaluation of Z depends on the evaluation of X and Y, the value of X and Y will have to be computed first before the computation of the value of Z. So, even if we use three processing elements, the third one will have to wait for the completion of the task of computation of X and Y first. Consequently, there will be no time saving at all.]

11. Construct a dataflow graph for a given program and evaluate the result.

Logic Program:

$o := (a + b) * (c - d)$,
$a := 4$,
$b := 2$,
$c := 3$,
$d := 1$.

[**Hints:** A dataflow graph for the given program is given here vide Fig. 1.28.

Given, $o := (a + b) * (c - d)$

$\therefore$ The result $= (4 + 2) * (3 - 1)$
$= 8 * 2$
$= 16$.

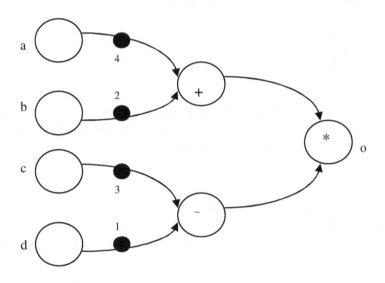

Fig. 1.28: A dataflow graph for the given program   ]

12. Show how concurrent resolution can be accomplished for the following logic
    programs. Mention where AND/OR/Stream parallelism is employed.

    a)
        Cl_1:Likes-mountaineering(X) ←Likes-adventure(X), Likes-snow(X),
                                          Likes-climbing(X).
        Cl_2: Likes-adventure(t) ←.
        Cl_3: Likes-snow(t) ←.
        Cl_4: Likes-climbing(t) ←.

    b)
        Cl_1: Likes-mountaineering(X) ←Likes-adventure(X), Likes-snow(X),
                                          Likes-climbing(X).
        Cl_2: Likes-adventure(a) ←.

        Cl_3: Likes-adventure(t) ←.
        Cl_4: Likes-snow(t) ←.
        Cl_5: Likes-climbing(t) ←.

    c)
        Cl_1: Int(1) ←.
        Cl_2: Int(N) ←Int(N-1).

Cl_3: Evaluate-fact(N) ←Int(N).
Cl_4: Evaluate-fact(N) ←N * Evaluate-fact(N-1).

[**Hints:**

a) AND-parallelism takes place when Cl_2, Cl_3, Cl_4 are resolved in parallel with Cl_1 during concurrent resolution.

b) OR-parallelism takes place when Cl_2 and Cl_3 are tried to resolve concurrently with Cl_1.

c) Stream-parallelism takes place when the factorial computation of say, (N-1) takes place with the generation of the integer N.]

References

1. Antoniou, G., *Nonmonotonic Reasoning*, MIT Press, Cambridge, MA, pp. 21-160, 1997.
2. Bender, E. A., *Mathematical Methods in Artificial Intelligence*, IEEE Computer Society Press, Los Alamitos, CA, chapter 1, pp. 26, 1996.
3. Besnard, P., *An Introduction to Default Logic*, Springer-Verlag, Berlin, 1989.
4. Buchanan, B. G. and Feigenbaum, E. A., "DENDRAL and Meta-DENDRAL: Their applications dimension," *Artificial Intelligence*, vol. 11, pp. 5-24, 1978.
5. Bundy, A., "Will it reach the top? Prediction in the mechanics world," *Artificial Intelligence*, vol. 10, pp. 129-146, 1978.
6. Clark, K. and Gregory, S., *PARLOG Parallel Programming in Logic*, Dep. Computing, Imperial College of Science and Tech., doctoral report, 1984/4, April 1984.
7. Chakraborty, A., *Cognitive Cybernetics- A Study of the Behavioural Models of Human-machine Interactions*, Ph. D. thesis, Jadavpur University, 2005.
8. Chakraborty, A., Konar, A., *Emotional Intelligence: A Cybernetic Approach*, Springer, Heidelberg, 2006.
9. Chakraborty, A., Sanyal, S. and Konar, A., "Semantic stability of logic," J. of Institute of Engineering (to appear).
10. Chimenti, D., Gamboa, R., Krishnamurthy, R., Naqvi, S., Tsur, S. and Zaniolo, C., "The LDL system prototype," *IEEE Trans. on Knowledge and Data Engg.*, vol.2, no. 1, March 1990.
11. Dougherty, E. R. and Giardina, C. R., *Mathematical Methods for AI and Autonomous Systems*, Prentice-Hall, Englewood Cliffs, NJ, 1988.
12. Halder, S., *On the Design of Efficient PROLOG Machines*, M.E. dissertation, Jadavpur University, Calcutta, India, 1992.

13. Hwang, K. and Briggs, F. A., *Computer Architecture and Parallel Processing*, McGraw-Hill, NY, 1986.
14. Jackson, P., *Introduction to Expert Systems*, Addision-Wesley Publishing Co., Great Britain, 1988.
15. Konar, A., *Artificial intelligence and Soft Computing: Behavioral and Cognitive Modeling of the Human Brain*, chapter 5, CRC Press, Boca Raton, Florida, 1999.
16. Konar, A., *Computational Intelligence: Principles, Techniques and Applications*, Springer, 2005.
17. Kuo, B. C., *Automatic Control Systems*, Prentice-Hall, Englewood Cliffs, NJ, 1975.
18. Kuo, B. C., *Digital Control Systems*, Holt-Saunders, Tokyo, 1980.
19. Lloyd, J. W., *Foundations of Logic Programming*, Springer-Verlag, NY, 1987.
20. Mark, S., *Introduction to Knowledge Systems*, Morgan Kaufmann, San Mateo, CA, 1995.
21. Mahanti, A. and Daniels, J. C., "A SIMD approach to parallel heuristic search," *Artificial Intelligence*, vol. 60, pp. 243-282, 1993.
22. McDermott, D. and Doyle, J., "Nonmonotonic Logic I," *Artificial Intelligence*, vol. 13(1-2), 1980.
23. Moore, R. C., Possible-world semantics for autoepistemic logic, in Readings on *Nonmonotonic Reasoning*, Ginsberg, M. (Ed.), Morgan Kauffmann, San Mateo, CA, pp. 137-142, 1987.
24. Murata, T., "Petri nets: Properties, Analysis and Applications," *IEEE Proc.*, vol. 77, no. 4, April 1989.
25. Nagrath, I. J. and Gopal, M., *Control Systems Engineering*, Wiley Eastern, New Delhi, 1983.
26. Newell, A. and Simon, H. A., "The Logic Theory Machine," *IRE Trans. on Information Theory*, 1956.
27. Nilson, N. J., *Principles of Artificial Intelligence*, Morgan Kaufmann, San Mateo, CA, 1995.
28. Nyquist, H., "Regeneration Theory," *Bel System Technical Journal*, vol. 11, pp. 126-147, 1932.
29. Oldfield, I. V., Mem, S., "Logic programs and an experimental architecture for their execution," *IEEE Proc.*, vol. 133, part E, pp. 163-167, May 1986.
30. Patterson, D. W., *Intro. to Artificial Intelligence and Expert Systems*, chapter 5, pp. 92-95, Prentice-Hall, Englewood Cliffs, NJ, 1990.
31. Reiter, R., "A logic for default reasoning," *Artificial Intelligence*, vol. 13, pp. 81-133, 1980.
32. Rich, E., *Artificial Intelligence*, McGraw-Hill, NY, 1983.
33. Shannon, C. E., "Automatic Chess Player," *Scientific American*, vol. 48, p. 182, 1950.
34. Shapiro, E. Y., *A Subset of Concurrent PROLOG and its Interpreter*, Tech. Report TR-003, Inst. for New Generation Comput. Technol., Jan 1983.

35. Shortliffe, E. H., *Computer-based Medical Consultations: MYCIN*, Elsevier, New York, 1976.

36. Tick, E. and Warren, D. H. D., "Toward a pipeline PROLOG processor," in *Proc. 1984 Int. Symp. on Logic Programming*, pp. 29-40, Feb. 1984.

37. Warren, D. H. D., *Implementing PROLOG- Compiling Predicate Logic Programs*, Res. Rep. 39 and 40, Department of Artificial Intelligence, Univ. of Edinburgh, 1977.

38. Yasuura, H. et al., "A hardware algorithm for unification on logic programming languages," *WGEC'84-67-2*, Inst. Electron. Inform. Commun. Engg., Japan, March 1985.

2

Parallel and Distributed Models for Logic Programming— A Review

The chapter provides a review of some well-known models of parallel and distributed logic programming. It begins with the well known RAP-WAM architecture and gradually explores the scope of parallelism in AND-OR logic program languages, CAM based PROLOG machines and many others. The latter part of the chapter provides a Petri net like framework for distributed reasoning using logic programs. The discussion on Petri net based models includes Murata's work and its extensions by Jefferey et al. The chapter comes to an end with a discussion on the scope of the book with special emphasis on concurrent resolution of logic program clauses using Petri nets.

2.1 Introduction

Classical models of logic programs employ SLD (Select Linear Definite clauses) resolution to execute the program in a sequential manner. Because of the sequential participation of the program clauses in SLD resolution tree, the time complexity of an SLD program is proportional to the number of program clauses. An examination of typical logic programs reveals that there exists ample scope of concurrently resolving a number of program clauses. Such concurrent resolution of program clauses can save significant computational time in the process of execution of a logic program.

The chapter explores the different types of parallelisms in a logic program and their possible implementation/realization by efficient hardware/software means.

In the last chapter we have examined AND-, OR- and Stream-parallelism and noted that random selection of AND/OR clauses in concurrent resolution may sometimes result in a conflict in the variable bindings. This conflict can be avoided by restricting unwanted concurrent resolution of AND-parallel or OR-parallel program clauses. The main emphasis of this chapter is to design proper control strategies to implement restricted AND/OR parallelism in concurrent resolution.

Several methods of AND-OR parallelism in logic programming languages have been addressed in this chapter. Takeuchi's work [15, 16, 17, 18], for instance, in

A. Bhattacharya et al.: *Parallel and Distributed Models for Logic Programming — A Review*, Studies in Computational Intelligence (SCI) **24**, 57–105 (2006)
www.springerlink.com   © Springer-Verlag Berlin Heidelberg 2006

this regard needs special mention. To implement restriction in the selection of program clauses for concurrent resolution, Takeuchi employed a guard part in the body of program clauses. This guard part of the clause helps avoiding unwanted resolution of AND/OR parallel program clauses.

Kale's AND-OR tree model [7] also provides a new approach to the solution of a query using a directed acyclic graph. This graph provides a framework for orderly selection of program clauses to answer a query. The advantage of Kale's method lies in its inherent parallelism and pipelining in the execution of a logic program.

Classical SLD resolution usually requires a stack like structure for its efficient execution. To enhance the speed of execution of program clauses, Naganuma et al. [12] suggest an alternative framework for execution of PROLOG programs using Content Addressable Memory (CAM) instead of a stack. The advantage of this CAM based machine includes an automatic realization on (i) argument unification of two predicates and (ii) removal of unused variable bindings. It is important to note that unused variable bindings is mandatory in concurrent resolution of clauses, and an automatic removal of this garbage ensures restricted AND/OR parallelism in the PROLOG program.

In a recent paper, Patt [13] provides a benchmark analysis of a set of typical PROLOG programs on standard or modified architectures of commercial machines. The theme of his analysis includes the levels of pipelined stages in the execution of program. Patt compiles a PROLOG program to WAM (Warren Abstract Machine) code and then provides options to run the code on a machine or to recompile it for execution on a commercial machine. The main advantage of his analysis is to determine an optimal sequence of execution of a PROLOG program to utilize parallelism at all four possible levels: (i) the language level, (ii) the compilation level, (iii) the processor implementation level and (iv) the system configuration level.

An alternative scheme to execute concurrent resolutions in a logic program is to use a Petri like net model. Murata [10, 14] has shown that all possible solution of a query in a logic program can be determined by time invariant solutions of a nonlinear equation $A \circ X = 0$, where A is an incidence matrix representing the structure of a Petri net and X is a solution vector containing the variable bindings for the desired goal. Various modifications of Murata's pioneering work in Petri net modeling in logic program are available in the current literature on machine intelligence. Jefferey and Murata's Petri net model [6] that allows deferred substitution of variables in the resolution of two other program clauses, in this regard, needs special mention.

The chapter examines all the above works with examples and discusses the scope of the book in view of the above works. It suggests further modification of Petri net models to realize many other forms of parallelism which remained unexplored in the above reviews. Potential parallelism in a PROLOG program in view of a RAP-WAM architecture is addressed in section 2.2. The mapping of a PROLOG program onto a parallel processing engine is outlined in section 2.3. The scope of parallelism in AND-OR logic programming languages is introduced in section 2.4. Kale's AND-OR tree model for logic programming is outlined in section 2.5. The CAM based architecture of PROLOG machines is outlined in section 2.6. Performance analysis of a PROLOG program on different machine architectures is presented in section 2.7. Section 2.8 introduces the scope of logic programming using Petri net. The scope of the book is discussed in section 2.9. Concluding remarks are appended in section 2.10.

2.2 The RAP-WAM Architecture

Yan [19] presented a detail survey on the scope of parallel realization of knowledge computing on multiprocessor architecture. A part of his survey focussed on the scope of parallel processing in both compilation phase and runtime phase of PROLOG programs. Yan stressed the needs for concurrent realization of PROLOG on a RAP-WAM machine. The WAM (**W**arren **A**bstract **M**achine) generates an intermediate code during the compilation phase of a PROLOG program that exhibits the fullest degree of parallelism in the program itself. The RAP (**R**estricted **AND P**arallelism), on the other hand, reduces the overhead associated with the run time management of variable binding conflict between goals. RAP includes a compile time analysis to identify the clauses entangled in the 'binding conflict problem' and determine the valuation space of the variables at run time level. This has significantly extended the WAM with the implementation of RAP by the above scheme.

Hermenegildo and Tick [5] proposed a new scheme for parallel realization of a PROLOG program on a multiprocessor architecture by employing the composite benefits of WAM and RAP machines. The design considerations proposed by them needs special mention, and are as follows:

(a)   Potential parallelism in a PROLOG program can be represented by a graph, where the take-off arcs from a vertex on the graph represents the parallel tasks in the program.

(b)   Goal stacks may be attached with each processing element (PE). When a parallel call is invoked, the concurrent tasks are mapped autonomously to the stacks of the less busy or idle processing elements.

(c)   Since the processing elements require non-uniform time to handle different tasks, a buffer is allocated to each PE to check the pending messages, if any, received from other PEs.

(d)   Synchronization and co-ordination of parallel call needs to be incorporated.

(e)　　Markers are employed to identify the point in the program at which backtracking should start. Hermenegildo [5] also described how the register contents of the PEs are appropriately saved before another clause is executed.

2.3 Automated Mapping of a Logic Program onto a Parallel Architecture

Ganguly, Silberschatz and Tsur [3] presented a new algorithm for automated mapping of given logic program onto a parallel processing architecture. Their entire work can be sub-divided into two major heads. First they presumed a given architecture with a fixed interconnection topology where the processors connected by a direct link can only communicate between them. Thus partitioning the program into modules that need to share variables are mapped onto adjacent processors. This is required to transfer the result of variable bindings by a processor to another that cannot proceed without the result of bindings from the former processor. The proposed architecture, thus, to some extent includes a pseudo-pipelining along with parallel processing.

In the latter part of their paper [3], Ganguly, Silberschatz and Tsur relaxed the restriction of communication between the adjacent processors only. They have shown that a significant speed up is possible by relaxing the above constraint, which however puts extra burden to on-line network management. However, they did not address the problem of network management in the above paper.

2.4 Parallel AND-OR Logic Programming Language

Takeuchi in his recent book on parallel logic programming [15] presented a new concept on realization of parallel AND-OR logic programs. Takeuchi et al. [17, 18] designed a new language AND OR-II that includes the complete realization of the proposed AND-OR parallelism. A schematic review of their work is outlined below.

According to Takeuchi et al. a world is defined by the conjunction of its atomic clauses. These atoms in a world are executed in parallel. This indirectly has correspondence with AND-parallelism. During the process of resolution of an OR-clause with similar clauses in the body of a program, a non-determinism appears as the world proliferates into different worlds. The naive implementation of this proliferation is to make a copy of a conjunction of all atoms, but this creates an extra overhead and thus is not acceptable. An alternative realization of OR-parallelism can be carried out using the *graph coloring scheme*.

To illustrate the coloring process let us consider the following logic program that includes both AND- and OR-parallelisms.

Program AND-OR parallelism

Compute (X, Z) ←Pick-up (X, Y), Square (Y, Y_2),
         Cube (Y, Y_3), Add (Y_2, Y_3, Z).   (2.1)

Square (X, Y) ←Y = X * X.   (2.2)

Cube (X, Y) ←Y = X * X * X.   (2.3)

Add (X, Y, Z) ←Z = X + Y.   (2.4)

Pick-up ([X|L], Y) ←Y = X.   (2.5)

Pick-up ([- | L], Y) ←Pick-up (L, Y).   (2.6)

In the above logic program pick-up is an OR-predicate, while Pick-up, Square, Cube, Add are AND-predicates. Let us consider the list L = [1, 2, 3]. The instantiation of X = 1, X = 2 and X = 3 should be done in the above program in sequence if there was no parallelism. However, when AND- and OR-parallelisms are allowed together, the parameter passing by the pickup OR-predicate to the evaluation process of square and cube should be done in pipeline, and then the addition process should be active. Thus we want to implement a three-stage pipeline with a parallel operation between computing square and cube. A schematic diagram of the overall system is presented in Fig. 2.1.

To implement the coloring scheme in Fig. 2.1, let us consider a notion of vectors, where the positions of the elements in the vector are time-tagged. Such

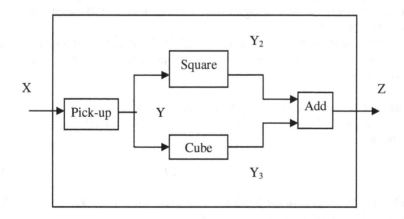

Fig. 2.1: The data flow graph of compute

time-tagging ensures true pipelining in the system. Alternatively, the elements of the vector may be colored and the addition operation should be executed on tokens of same color. Thus assigning a vector $X = [1, 2, 3]$ we find three colored tokens in vector form at Y, given by

$$Y = [\, v \, (1, \alpha_1), \, v \, (2, \alpha_2), \, v \, (3, \alpha_3) \,] \tag{2.7}$$

where α_1, α_2 and α_3 denote three colors. The Y_2 and Y_3 vectors thus take the form of

$$Y_2 = [\, v \, (\, 1, \alpha_1), \, v \, (\, 4, \alpha_2), \, v \, (\, 9, \alpha_3) \,], \tag{2.8}$$
and
$$Y_3 = [\, v \, (\, 1, \alpha_1), \, v \, (\, 8, \alpha_2), \, v \, (27, \alpha_3) \,] \tag{2.9}$$

and consequently Z takes the form

$$Z = [\, v \, (\, 2, \alpha_1), \, v \, (\, 12, \alpha_2), \, v \, (\, 36, \alpha_3) \,]. \tag{2.10}$$

Takeuchi [16] in his work on AND-OR parallel language and its realization presented the operational semantics of parallel computations in a logic program. As already discussed earlier, the conjunction of atoms together defines the world of the program. Thus, replacement of a body clause by two or more conjunctive atoms is called *proliferation into new worlds*. Determining the sequence of execution of the goal clauses, however, is a crucial issue in controlling the concurrency in a logic program. The two semantic rules containing guard clauses can, however, be employed to handle these problems.

According to Takeuchi two semantic rules, called the rule of suspension and the rule of commitment for controlling the concurrency in AND-OR -parallelism are discussed in this section.

In order to understand the semantics of AND-OR parallel computation in a logic program, following Takeuchi we define the syntax of guarded and non-guarded clauses.

Definition 2.1: *A **guarded clause** can be represented by a head G_o, a guard part $G_1, \ldots\ldots, G_n$ and a body $B_1, \ldots\ldots, B_m$. Thus formally, a guarded clause takes the following form:*

$$G_o \leftarrow G_1, G_2, \ldots\ldots\ldots, G_n. \, / \, B_1, B_2, \ldots\ldots, B_m. \tag{2.11}$$

Definition 2.2: *A **non-guarded clause** comprises of a head G_o and a body consisting of literals like $B_1, B_2, \ldots\ldots, B_m$. Thus,*

$$G_o \leftarrow B_1, B_2, \ldots\ldots\ldots, B_m. \tag{2.12}$$

An atom in the body part in both a guarded and a non-guarded clause can be an AND-predicate and an OR-predicate.

In AND-parallelism, atoms in the body clause that together forms a world are executed in parallel. On the contrary, in OR-parallelism, the predicates in the body being the OR clauses, a given clause can give rise to a number of new clauses by unifying the body OR-clauses of a goal clause with several heads of other clauses. When a goal clause is invoked, the parent world corresponding to the goal clause proliferates into several new worlds following the resolution of new clauses with the body clauses of the goal. This is usually referred to as OR-parallelism.

In case there are more than one way to resolve a goal clause, some restrictions are imposed on the selection of the clauses that can satisfy the goal. For instance, let us consider a goal clause G that calls a guarded AND clause C.

Formally,

$$G \leftarrow C.$$

$$C \leftarrow G_1, G_2, \ldots\ldots\ldots, G_n. \mid B_1, B_2, \ldots\ldots\ldots, Bm. \tag{2.13}$$

where the notations in the above clauses have their usual meaning.

In connection with realization of AND and OR parallel computation of a logic program Takeuchi presents two rules for both AND- and OR-operational semantics.

Operational semantics for AND-predicates

This can be best represented by the *rule of suspension* and the *rule of commitment* introduced as follows.

Let us consider a logic program comprising of clauses like

$$G \leftarrow C.$$
$$C \leftarrow G_1, G_2, \ldots\ldots, G_n. \mid B_1, B_2, \ldots\ldots, B_m. \tag{2.14}$$

where G = head atom, C = guarded clause, $G_1, G_2, \ldots\ldots, G_n$ = guard part and $B_1, B_2, \ldots\ldots, B_m$ = body part.

Rule of suspension: To describe this rule we first define the term: *Guard computation*. 'Guard computation' of a clause C stands for both the head (G) unification and execution of the guard part, $(G_1, G_2, ..., G_n)$ in the last example program (2.14).

The rule of suspension states that *the clause C that includes a guard computation should not instantiate the head G, even if the guard part G_1, G_2, ...,G_n can be instantiated directly or indirectly by other clauses. Thus the goal unification process for the above set of rules is suspended.*

Rule of commitment: The rule of commitment states that *in case there exists no other rules that can instantiate the head G, then the clause C should be committed, provided its guard part is satisfied through unification with other clauses.*

The rule of commitment should always supercede the rule of suspension.

Operational semantics for OR-predicates

The operational semantics of OR-predicates must obey the following two rules.

Rule of suspension:

Consider two rules

$$G \leftarrow C.$$
$$C \leftarrow B_1, B_2, ..., B_m. \qquad\qquad (2.15)$$

where the second rule contains only non-guarded body clauses. The rule of suspension states that *the second rule should not satisfy the head atom G unless there is no other rules that can satisfy the head G. Thus the unification process of the head G is suspended.*

Rule of proliferation: *In case there exists N number of non-guarded clauses that succeed in unifying the goal G, then all the rules should be used for goal unification in parallel.*

For instance consider the set of rules:

$$
\left.
\begin{array}{lcl}
G & \leftarrow & C. \\
C & \leftarrow & B_1, B_2, ..., B_m. \\
C & \leftarrow & C_1, C_2, ..., C_m. \\
\cdot & & \\
\cdot & & \\
C & \leftarrow & D_1, D_2, ..., D_m.
\end{array}
\right\} \quad \text{N rules} \qquad (2.16)
$$

In the present example the world represented by the clause C of the rule 'G ←C.' should be proliferated by the N set of rules after the head unification is over.

2.5 Kale's AND-OR Tree Model for Logic Programming

Kale in one of his recent book chapters [7] presented an alternative formulation of problem solving using parallel AND-OR trees. The special feature of his scheme lies in ordering the search process for the OR-nodes that together constitutes the AND-node. The scheme of Kale is briefly outlined below. The following definitions are in order to explain the characteristic features of Kale's model.

Definition 2.3: *An **AND-OR tree** is a tree rooted with AND-nodes. The root node usually denotes a query of the form*

$$\leftarrow G_1, G_2, \ldots\ldots, G_n.$$

Each literal G_i, for i = 1 to n under the root node is written separately at the next level of the tree describing the queries

$$\leftarrow G_i. \text{ for } i = 1 \text{ to } n.$$

Each G_i is called an OR-node. OR-nodes have a single child AND-nodes that satisfy the following characteristics.
Let a Head $\leftarrow Q_i$ for i = 1 to n be an AND–node at depth 2 of the AND-OR tree under the node G_i, such that the Head uses the same literal as G_i but different terms, having possible variable bindings. Thus, on resolution of

$$\leftarrow G_i.$$

and

$$Head \leftarrow Q_i.$$

we find $\leftarrow Q_i [. /.].$

where [. /.] denotes the results of substitution of the variables of G_i by the terms in the Head. The substitution [. /.] is attached with the arcs connected from the OR-nodes to their child.
When $Q_i = \phi$, the corresponding AND-node is called a leaf. Such clause C where C = Head $\leftarrow Q_i$ reduces to Head $\leftarrow$, which is a fact.

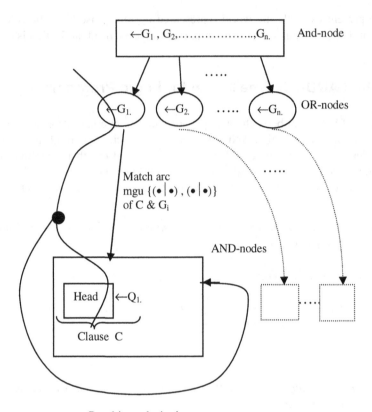

Resulting substitution

The AND-node: Head $\leftarrow Q_1$ at depth 2 of the AND-OR tree is
resolved with clause $\leftarrow G_1$ of depth 1, with the resolvent $\leftarrow Q_1[. | .]$

Fig. 2.2: An illustrative AND-OR tree

Figure 2.2 describes a typical AND-OR tree rooted with $\leftarrow G_1, G_2, \ldots\ldots, G_n$
and only one AND-node at depth 2 for illustration purpose.

Definition 2.4: *A **candidate solution-tree** for a literal G or for a query Q is a
sub-graph of the AND-OR tree for G or Q that satisfies the following constraints:*

The graph should include the root of the AND-OR tree.

*In case it includes an AND-node A, then it should also include the children
nodes of node A.*

In case it includes an OR node O, it should also include exactly one child node of node O.

Definition 2.5: *This arc connected from an OR-node to an AND-node is called a* ***match arc***. *This arc is labeled with the most general unifier (mgu), to be defined later(in chapter 3), obtained through the process of unification of a clause $\leftarrow G_i$ and the head part of the clause*

$$\boxed{\text{Head}} \quad \leftarrow Q_1 \text{ located at the OR-node and its AND-child respectively.}$$

Definition 2.6: *A* ***consistent solution-tree*** *is a candidate solution tree so that the labels attached with the matched arcs have a common unifying composition.*

Definition 2.7: *A* ***query*** *Q is called* ***solved*** *if the AND-OR tree having a root at Q has a consistent solution-tree.*

Definition 2.8: *Let θ_i be the mgu labeled with the match arc i. Assuming that there exist n number of OR-nodes in the solution-tree, we define a set S as*

$$S = \bigcap_{i=1}^{n} \theta_i$$

The projection of S on the variables of the query Q is called a ***solution*** *to Q.*

Example 2.1: Let us consider a problem of finding the value of variable A, such that the number A is both prime and belongs to the fibonacci series. There are two ways to solve the problem. First the problem can be subdivided into two heads, so that Prime (A) and Fibonacci (A) for A = 233, say, can be checked in parallel. A resulting truth value of the predicate

Fibonacci-and-Prime (233)

will be found only when the solution to Prime (233) and Fibonacci (233) both exist.

An alternative analogous problem may be to find the possible solutions of

Fibonacci-and-Prime (A).

where A is within a given range, say $0 < A < 500$. One way to solve this problem efficiently is to generate Fib(N, X), where X is the N-th Fibonacci number and then the result X may be passed on to the predicate Prime (X) for testing of X for a prime number. This scheme is similar to pipelining of two processes. A question then naturally arises: is there any definite rule for the judicious ordering of the predicates to be placed in the pipeline? For instance, should we test Prime (X) prior to Fib (N, X)? Obviously, this is a wrong choice, as most of the prime numbers do not belong to the fibonacci series.

The problem is exaggerated further when the body of the clause includes a number of predicates. For example, consider the problem of selection of a venue at a given date by two friends. This can be formally represented by the following clause

$$\text{Dinner-date (I, Y, Restaurant, Day)} \leftarrow \text{Likes (I, Y), Free (Y, Day), Enjoys (Y, Restaurant), Open (Restaurant, Day).} \quad (2.17)$$

Here, the problem is manifold. First we have to identify person "Y" whom "I" likes. This may result in many solutions. Then for each Y we must check whether Y is free on a given date and then we need to identify the restaurant that Y likes and only after that we must check whether the restaurant is open on that date. Thus if we maintain a strict pipelining of the predicates Likes, Free, Enjoys and Open in order, then perhaps we can find a time-efficient solution for the problem easily. Kale introduces a new representation of the partial ordering of the predicates by a graph where the ordering stands for sequencing the predicates in the way they will give rise to solutions. Such graphs are usually called Data Join Graphs (DJG).

A formal definition of DJG is presented as follows.

Definition 2.9: *A **DJG** is a directed acyclic graph where the nodes denote the ordering of events, here the solutions of predicates and the directed arcs connected between any two nodes are labeled with a predicate. Each DJG has a single start node and a finish node. The parallel activities (finding the solution of more than one predicate) are denoted by parallel arcs between two nodes of a DJG.*

Example 2.2: In this example we describe the DJG of the following query (vide Fig. 2.3):

$$\text{Query 1:} \leftarrow \text{Fib (N, F), Perf (P), X is F + P, Prime (X).} \quad (2.18)$$

The query 1 calls for finding the solutions X such that X = F + P, where F is the N-th fibonacci number and P is any perfect number.

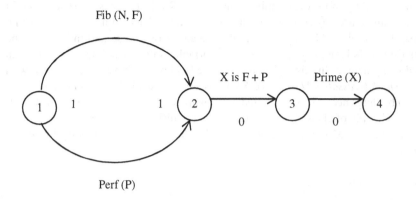

Fib (N, F)

Perf (P)

Fig. 2.3: The DJG corresponding to the given query

For example F = 1, 5 are two fibonacci numbers while P = 6 is a perfect number. Under this case, we have two solutions for X = (1 + 6) = 7 and X = (5 + 6) = 11. In this example computation of Fib (N, F) and testing of Perf (P) can be done concurrently. So the DJG has two arcs connected between nodes 1 and 2. Finding an efficient DJG for a given query, itself is a complex problem, and no formal solution to such problems is known till date.

2.6 CAM-based Architecture for a PROLOG Machine

In the process of SLD resolution, the AND-clauses are selected and unified with the heads of other clauses in sequence. For example, let us consider an AND-clause:

$$P (X, Y) \leftarrow Q (Y, Z), R (X, Y). \tag{2.19}$$

where Q and R are selected for variable unification in sequence. The process of involving an AND-literal for search in the head of existing clauses and its unification with a selected clause are two major steps in the resolution process. The first step may, hereafter, be referred to as *clause invocation*, while the latter may be described as *argument unification*. Unlike conventional Prolog, Naganuma et al. [12] considered the scope of argument unification of AND-clause Q (Y, Z) in parallel with the clause invocation of the AND-literal R (X, Y). Such time overlapping between these two processes is beneficial for increasing the computational time-efficiency of a logic program. Fortunately, the work presented in [12] employs this principle, thereby speeding up the process of inferential reasoning to as high as 100 KLIPS (Kilo Logical Inferences Per Second). Another interesting feature of their work lies in automatic re-instantiation and checking of the bound variables of Q as possible bindings of R. Let Y = α, Z = β be the resulting binding of Q. Under this circumstance, R (X, α) needs to be searched in the existing clauses. If no matched literal R (X, α) is detected in the head of some clause, the previous bindings of literal Q becomes of no use and thus should be destroyed. This is usually realized with a variable binding stack in a conventional PROLOG machine. In this reference [12], the binding stack has been realized with a *Content Addressable Memory* (CAM) instead of a stack. The destruction of the unused bindings here has been implemented with an automatic clearing of the CAM.

The principle of clause invocation, argument unification and removal of unused bindings in a CAM is presented below with reference to the following PROLOG program:

PROLOG Program:

←P (X, Y).	(2.20)
P (U, V) ←Q (U, V, W), R (W).	(2.21)
Q (a, b, c) ←.	(2.22)
Q (d, e, f) ←.	(2.23)
R (f) ←.	(2.24)

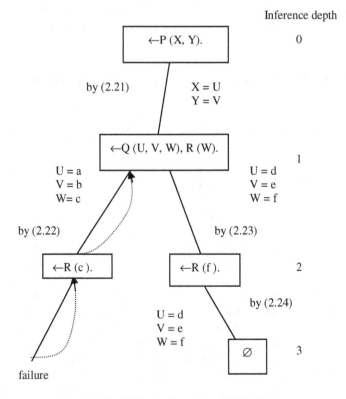

Fig. 2.4: The SLD-tree for a given PROLOG program

In the aforementioned program '←P (X, Y).' is the query, and the remaining clauses (2.21) to (2.24) constitute the logic program. Figure 2.4 describes the execution process of the above program by an SLD tree. The phrase: *Inference depth,* which will be used in our subsequent discussion is now informally defined.

Inference depth denotes the depth of nodes in an SLD tree. The root node, which corresponds to a query, has an inference depth zero. All other nodes in the tree that describe inferred clauses have an inference depth equal to the depth of the respective nodes counted from the root in the SLD tree. Here, the goal clause ←P (X,Y) is placed at depth zero, and it is resolved with the clause (2.21) to yield a new clause ←Q (U, V, W), R (W). The variables X and Y now have the bindings X = U and Y = V. The resulting clause thus obtained occupies an inference depth one in the given tree. This clause is then further resolved with clause (2.22) to yield ←R(c) and the variable values U, V, W obtained the bindings: U = a, V = b and W = c. The clause ←R (c) occupies a depth two in the SLD tree. The clause R (c) is now searched in the heads of the existing set of clauses, but unfortunately there is no such atomic clause in the existing heads, and consequently the resolution fails with a backtracking to the clause at depth one. The SLD-tree is then further expanded by invoking clause (2.23) for resolution with the clause at inference depth one to yield ←R (f). Finally, the clause (2.24) is resolved with the resulting clause ←R (f) of depth 2 to yield a null clause that occupies a depth 3 in the SLD tree. The bound value of variables thus obtained are X = U = d and Y = V = e.

The CAM employed to handle the above problems in an SLD tree includes five distinct fields. The first field denotes the inference depth of the clauses. It has a minimum value 1 as the first inferred clause occupies an inference depth 1. The second field of the CAM denotes variables present in the parent clause of the node under consideration. The third field denotes the depth of the parent clause containing the variables listed in field 2. The fourth field of the CAM represents the bound value of the variables considered in field 2. The fifth field of the CAM describes the depth of the clause that includes the resulting bindings of the variables present in field 4.

The execution of the given logic program starts with an empty CAM. After first resolution, the clause listed at inference depth 1 is generated. Thus field 1 of CAM (vide Fig. 2.5(a)) contains 1. The parent clause here being ←P (X, Y), the variable X, Y are inserted in the second field of the CAM. Since the depth of ←P (X, Y) is zero, zeroes are entered in the third field. The fourth field includes the bound values: U and V for the variables X and Y respectively. The fifth field denotes the inference depth of the clause containing the bound values U and V. Thus it is affixed with one.

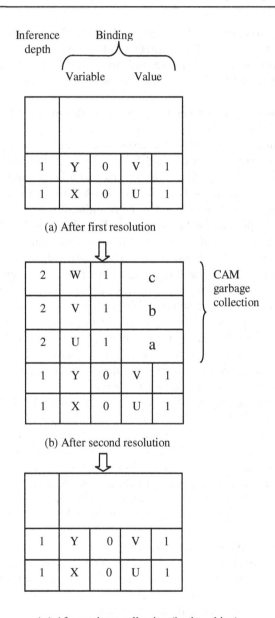

(a) After first resolution

(b) After second resolution

(c) After garbage collection (backtracking)

Fig. 2.5: CAM-based backtracking without trial stacks

After the second resolution, the CAM obtained in step (a) is expanded with three more rows. It should be mentioned here that once constant bindings are obtained, no further declaration of the depth of the clauses containing those bindings is needed. Thus the last two fields are merged into a single field describing the values of the variables only. Since the resulting clause after the second resolution is ←R (c), it has an inference depth 2, and thus the first field is now filled in with 2 (vide Fig. 2.5(b)). Further, as we have three variables U, V and W, three rows are employed to describe the variable bindings and other necessary information. The second field thus is filled in with variable names U, V and W present in the parent clause ←Q (U, V, W), R (W). Since the depth of the above clause containing U, V and W is one, the third fields are filled in with one. Further, the resulting bindings of U, V and W are a, b and c respectively, and thus they are stored in the fourth field of the CAM. It may be noted that after ←R (c), the resolution process cannot proceed further, and consequently we need to backtrack to the node at depth 1 of the SLD-tree. Since the bindings obtained in the left-side of the clause ←Q (U, V, W), R (W) are no longer useful, in case of a stack realization two consecutive POPs could destroy the garbage bindings. This has been represented in the present CAM-based architecture by removal of all the top three rows (vide Fig. 2.5(c)). The storage of the CAM in the subsequent two steps can now easily be visualized and thus is not included in the present discussion.

2.7 Performance Analysis of PROLOG Programs on Different Machine Architectures

The work presented in [4] addresses various alternative models for execution of a PROLOG program. Performance of a PROLOG program can be improved by detecting and analysing parallelisms at different levels. These levels include (i) language level, (ii) compilation level, (iii) processor implementation level and (iv) system configuration level. The language level is concerned with detection of four typical forms of parallelisms namely (i) AND-parallelism, (ii) OR-parallelism, (iii) Stream-parallelism and (iv) Unification-parallelism. All these parallelism have already been discussed in section 1.6.1. At compilation level a data dependency graph may be employed to detect the independence of the intermediate machine codes, and thereby allowing them to be mapped to different hardwared units subsequently. Backtracking, that needs the control to trace back to the previous step, can also be analyzed in the compilation phase. There exist ample evidence of analysing data-dependency and backtracking in a PROLOG program vide [2]. The processor level parallelism deals with execution of the machine codes on different hardwared units in parallel within the processor. The system level parallelism on the other hand employs a multiprocessing system with a number of processors to take care of the data flow among the processors through an interconnection network. A parallel program is mapped onto the distributed processors of a multiprocessor system to enhance the speed of execution of a PROLOG program.

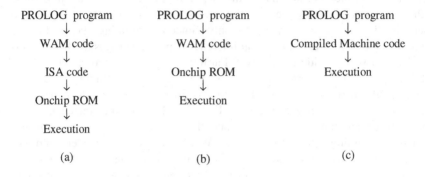

Fig. 2.6: Three different levels of organization of PROLOG programs

Patt [13] provides a bench mark analysis of a set of typical PROLOG programs on standard/modified architectures on popular commercial machines. The basic theme of his analysis lies in determining the levels of pipelined stages in the process of execution of a PROLOG program. His analysis is restricted to an uniprocessor architecture only. Patt compiles a given PROLOG program to Warren Abstract Machine (WAM) code and then selects options whether to run the WAM code directly on a machine or to recompile it to the machine code of commercially available machine for execution. While recompiling the WAM code Patt prefers the well known ISA code for machine implementation. Thus three typical levels of organisations of a PROLOG program emerge from his analysis. These are schematically described in Fig. 2.6.

While experimenting on NCR/32 machine at the University of California, Berkeley Campus Patt noted that the first mode (Fig. 2.6(a)) exhibited a poor performance in comparison to the others.

2.8 Logic Programming Using Petri Nets

Coined after the seminal work of Karl Adam Petri, the phrase *Petri nets* has already proved itself successful in mathematical modeling and simulation of various systems. Some important application of Petri nets include (a) representation of programs for dataflow computing, (b) deadlock avoidance in operating systems, (c) Time-scheduling of discrete event systems, (d) protocol management in a communication system, (e) representation of a context sensitive language, and (f) resource sharing in a multi-process system. Unfortunately, the use of Petri nets in knowledge engineering in general and in logic programming in particular has started only in the last decade. This section highlights the significance of Petri nets in efficient reasoning with logic programs.

A Petri net is a directed bipartite graph comprising of two types of nodes: *places* and *transitions* and directed arcs to represent connectivity from places to transitions and vice-versa. The importance of Petri nets in knowledge engineering arises because of the distributed organization of its structure, capable of holding the smallest fragments of a program clause onto several components of its structure. Such fragmentation of program resources onto smaller components of a Petri net is needed for two reasons. First all possible parallelism in a program can be fully exploited because of modular organization of the program onto several structural units of a Petri net. Secondly, the fragmentation of the program helps in protecting parts of the program, instead of destruction of the entire program due to hardware failure.

Besides, Petri nets when used in logic programming offers some additional benefits. A Petri net graph, for instance, is an ideal choice of representing *structural pipelining* of program data resources [8]. The firing sequence of transitions, as if, denote the firing sequence of the *modus ponens* rules and the derived inferences, thus, traverse through the net following the firing sequence of transitions. Further, because of the structural benefits of Petri net models, all typical parallelisms of a logic program, such as AND-, OR-, Stream- and Unification-parallelisms can be realized on the proposed structure. The provisions for concurrent firing of transitions in a Petri net facilitates the users with the additional benefits of resolving multiple program clauses in bunches on isolated transitions. The book, in fact, emphasizes these issues in more detail in the subsequent chapters.

Murata first examined the scope of reasoning in logic programs using Petri nets. In one of his early papers [10, 14], he took a bold attempt to represent a logic program by a set of state equations, whose time-invariant solutions provide answers to the users' query. His model can be best introduced with example 2.3.

Example 2.3: Consider the following logic program comprising the following five Horn clauses, the last clause being the query.

Logic Program:

$$
(1) \text{ Parent (david, mary)} \leftarrow. \tag{2.25}
$$

$$
(2) \text{ Parent (mary, tom)} \leftarrow. \tag{2.26}
$$

$$
(3) \text{ Ancestor (X, Y)} \leftarrow \text{Parent (X, Y).} \tag{2.27}
$$

$$
(4) \text{ Ancestor (X, Z)} \leftarrow \text{Parent (X, Y), Ancestor (Y, Z).} \tag{2.28}
$$

$$
(5) \leftarrow \text{Ancestor (X, tom).} \tag{2.29}
$$

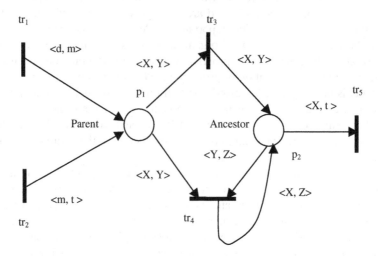

Fig. 2.7: Petri net corresponding to the given logic program

The generic state equation of a logic program realized with a Petri net (vide Fig. 2.7) is given by

$$A \circ X = O \qquad\qquad (2.30)$$

where $A = [a_{ij}]$ is an incidence matrix, X provides a solution to the users' query and o denotes a matrix product with substitution. For construction of the incidence matrix A, Murata employed the following principles:

Given a logic program consisting of n clauses and m distinct predicate symbols, the $(n \times m)$ incidence matrix A of a high-level net corresponding to the logic program can be obtained by invoking the following procedure.

- *Each clause in the program will be one row of the matrix (one transition in the net).*

- *Each distinct predicate symbol in the program will be one column of the matrix (one place of the net).*

- *The (i, j)th entry a_{ij} is the argument of the i-th clause and in the j-th predicate symbol, where an argument to the right of the $\leftarrow$ is prefixed with a negative sign. If the j-th predicate symbol appears more than once in the i-th clause, then a_{ij} will be the formal sum of all those arguments in the i-th row and j-th column.*

The procedure presented above converts the given logic program into the incidence matrix as follows:

$$A = \begin{array}{c} tr_1 \\ tr_2 \\ tr_3 \\ tr_4 \\ tr_5 \end{array} \begin{array}{cc} \text{Parent}(p_1) & \text{Ancestor}(p_2) \\ \langle d, m\rangle & 0 \\ \langle m, t\rangle & 0 \\ {-}\langle X, Y\rangle & \langle X, Y\rangle \\ {-}\langle X, Y\rangle & {-}\langle Y, Z\rangle + \langle X, Z\rangle \\ 0 & {-}\langle X, t\rangle \end{array}$$

where d, m, t denotes David, Mary and Tom respectively.

$$X_1 = \begin{array}{c} tr_1 \\ tr_2 \\ tr_3 \\ tr_4 \\ tr_5 \end{array} \begin{pmatrix} \varnothing \\ \{\} \\ \{m \mid X, t \mid Y\} \\ \varnothing \\ \{m \mid X\} \end{pmatrix}$$

$$X_2 = \begin{array}{c} tr_1 \\ tr_2 \\ tr_3 \\ tr_4 \\ tr_5 \end{array} \begin{pmatrix} \{\} \\ \{\} \\ \{m \mid X, t \mid Y\} \\ \{d \mid X, m \mid Y, t \mid Z\} \\ \{d \mid X\} \end{pmatrix}$$

where $\varnothing$ denotes no firings and $\{\ \}$ denotes a firing with no substitutions. The above vectors can be interpreted as "T-invariants" of the high-level net since they satisfy $A^T o X_1 = 0$ and $A^T o X_2 = 0$, where o denotes "matrix-product with substitutions."

Murata and Yamaguchi [11] in early 1990s presented an alternative approach to automatic reasoning using Petri nets. They devised a new model of Petri net for handling the forward and backward reasoning problems that supports the resolution theorem under the framework of classical logic. The work reported in the work [11] is primarily based on the following principles.

- *A general program clause, containing one or more literals in both the body and the head part, can be denoted by a Petri net with a number of input places equal to the number of antecedent literals and a number of output places equal to the number of consequent literals.*

- *Resolution of two or more clauses has been symbolized by firing of a transition.*

- *On firing of a transition the tokens possessed by the input and output places of the fired transition are updated by the following principle.*

For a propositional logic based programs, the derived token M'(.) (markings in Petri net terminology) at input place p and output place q of a fired transition can be expressed as a function of its value M(.) before firing:

$$M'(p) = (M(p) + \{-1\}) \cup M(p) \qquad (2.31)$$

$$M'(q) = (M(q) + \{+1\}) \cup M(q) \qquad (2.32)$$

The tokens at all other places r ($\neq p$ or q) remain unchanged.

Murata et al. extended the above model for automated reasoning with predicate logic based programs. The detailed discussion on the model goes outside the scope of the book, and is omitted.

Li [9] demonstrated a new approach to automated reasoning in Logic program for both Horn and non-Horn clauses with negation in the body of the clause. This is an uniform approach that is applicable jointly to monotonic and non-monotonic systems with negated literals in the body of the clauses.

Jeffrey et al. [6] presented an alternative scheme for goal directed reasoning in a Horn clause based logic program. The representation scheme proposed by them significantly differs with the currently available models of Petri nets. For instance, a typical Horn Clause 'R (X, U) ←P (X, Y), Q (U, V).' following Jeffrey et al. can be represented as shown in the Fig. 2.8.

It may be noted in Fig. 2.8 that, unlike conventional Petri net representation, the direction of the arcs are reversed. The following aspects of the reasoning undertaken by Jeffrey et al. need special mention.

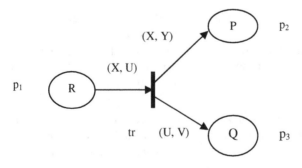

Fig. 2.8: Petri net representation of a Horn clause as advocated by Jeffrey et al.

The instantiation space of the variables in most Petri net based logic programs are determined locally after taking into consideration the binding of the variables of all arc functions associated with a transition. One major drawback of such a policy of instantiation lies in long deferred substitution of variables until the bound tokens propagate to the associated places of the same transition after several firings. Jeffrey's scheme however handles the situation in a faster and robust approach. In absence of proper constant bindings, they replace the variable component of tokens in a place by a renamed variable [1] and later substitute it by the value that is attained at some other places associated with the same transition. Consequently the variable components of token gets updated by a revised renamed variable until a constant binding of the same variable is attained transitively in a long chain of transitions.

Jeffrey et al. described the SLD resolution of program clauses by transformation of markings in the places. The updated tokens in the places usually are renamed variables or constants. In the process of updating of tokens in the places, the renamed variables in the token are transformed to constants. When all the variables/ renamed variables in the token of a place are replaced by constants, a goal or sub-goal may be obtained, and the place becomes empty. Consequently, when the (renamed) variables at all places become constants, all places become empty, and no further results can be derived from the Petri net. It may be added here that an empty place may regain tokens due to firing of a transition connected to the place.

Example 2.4: This example illustrates the principle of Jeffrey and Murata's model outlined above with a typical logic program. Consider, for instance the following logic program:

$\leftarrow t_c\ (a, Y).$	$(\equiv Q1)$	(2.33)
$t_c(X, Y) \leftarrow r(X, Y).$	(tr_1)	(2.34)
$t_c\ (X, Y) \leftarrow r(X, Z), t_c\ (\,Z, Y).$	(tr_2)	(2.35)
$r(a, b) \leftarrow.$	(tr_3)	(2.36)
$r(b, c) \leftarrow.$	(tr_4)	(2.37)
$r(a, d) \leftarrow.$	(tr_5)	(2.38)

Figure 2.9 describes the given logic program and the query $\leftarrow t_c\ (a, Y)$. One possible firing sequence of the transitions that leads to a successful evaluation of the query is presented in Fig. 2.10. Let the arc function $<X, Y>$ present in the input arc of transition tr_1 has a possible instance of solution $X = X_1$ and $Y = Y_1$. Thus after resolution of the program clause (2. 34) with the query (2. 33), X_1 and Y_1 are instantiated with one pair of new value and the resulting set is given by $\{X_1/a,\ Y_1/Y\}$. Consequently, a new token $<a, Y>$ arrives at the place corresponding to the predicate r. Transition tr_3 is then fired and a new instantiation of $Y = b$ is obtained.

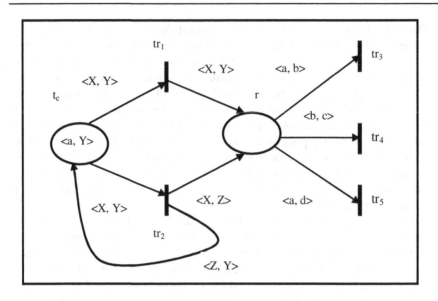

Fig. 2.9: Representation of the queried program by a Petri net

According to Jeffrey et al. [6], since a complete solution $<X_1, Y_1>$ now has been obtained the places are kept free from tokens to allow subsequent firing of other transitions for generating new solutions of the given query. Other solutions of the query $\leftarrow t_c$ (a, Y) are not presented here for lack of space; rather a case of failure is demonstrated vide Fig. 2.11. In this figure, the topmost configuration describes a initial situation in the Petri net. On firing of transition tr_5, Z_1 is instantiated with d, by matching $<a, Z_1>$ for place r with the input arc function $<a, d>$ of transition tr_5.

The value of $Z_1 = d$ is now updated in all places containing the variable Z_1. It is indeed important to note that in Jeffrey's model the variables in places correspond to a global variable in the entire logic program, in contrast to conventional Petri net models where variables are defined locally with reference to transitions.

The second module in Fig. 2.11 describes the resulting situation in the Petri net after firing of the transition tr_5. The transition tr_1 is now fired, and the second alternative instance $<X, Y> = <X_1, Y_1>$ is now presumed. The input arc function $<X_1, Y_1>$ of transition tr_1 thus constructed is now unified with $<d, Y>$ of place t_c and the resulting set of substitution obtained is given by $\{X_1/d, Y_1/Y\}$. This token value $X_1 = d$ and $Y_1 = Y$ is now inserted in place r following the guiding output arc function of transition tr_1. The token $<d, Y>$ is now matched with all the input arc functions of transitions tr_3, tr_4 and tr_5 respectively, but no possible variable bindings could be derived in the present context. This has been referred to as a *failing computation* in the work undertaken by Jeffrey et al. [6].

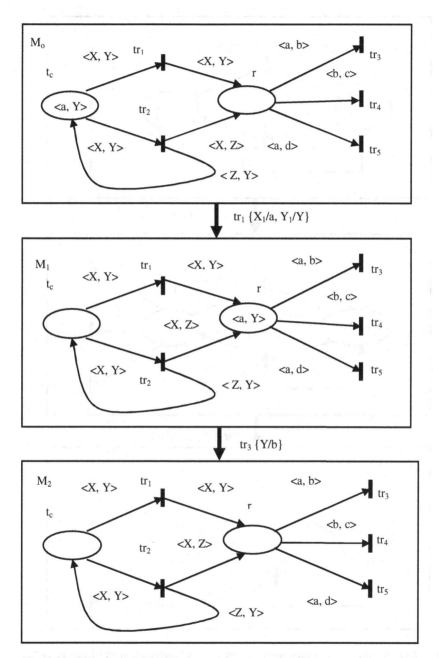

Fig. 2.10: A successful firing sequence Mo→M₁→M₂ of the given program. The arcs between frames are labeled with the fired transition and the variable substitutions made on firing of a transition

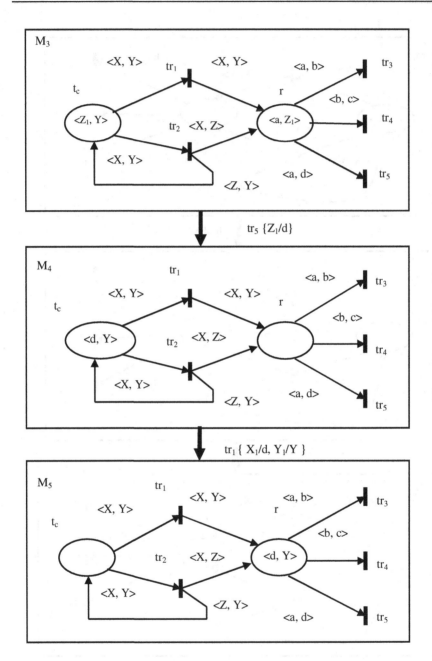

Fig. 2.11: A failing computation for the given queried Program through frames $M_0 \rightarrow M_3 \rightarrow$ $M_4 \rightarrow M_5$. M_0 to M_3 transition is not shown in the figure for clarity. It is to be noted that after firing of transition t_5, the variable Z_1 in tokens at both places r and t_c are updated in frame M_4

2.9 Scope of the Book

Reasoning in typical logic programming languages such as PROLOG or DATALOG is usually accomplished by automatically expanding an SLD-tree in the runtime phase of the program execution. Construction process of the SLD-tree involves sequential resolution of each pair of program clauses, and the resultant clause thus derived is passed on for resolution with a third program clause, judiciously selected from the program based on certain prerequisites for resolution. The process is continued until all possible solutions to the given query are evaluated. The SLD-tree building algorithm requires a linear order of time proportional to the number of program clauses that participate in the resolution process. Logic programs used in commercial/industrial applications generally include as many as 10,000 clauses. The computational time of an SLD-tree building algorithm thus been highly expensive prohibits the scope of logic programming in real time applications. Fortunately, the resolution principle in logic programming supports various types of parallelism. An appropriate realization of these parallelisms on an architecture, thus, may provide a new avenue to the aforementioned practical problem.

The objective of the book is to design a high-speed computational engine for logic programming that is capable of utilizing all possible parallelisms in the program. For a suitable implementation of the parallel processing inference engine, the book adopted a specialized data structure that can represent and reason with the program clauses by fully exploiting AND-, OR-, Stream- and Unification-parallelisms in the program. The data structure employed in the present context is a Petri net that besides having the above benefits supports the distributed organization of the program resources, such as predicates and their variable and constant arguments (here after called tokens) onto different modules of the network. Such fragmentation of the program into minute resources enhances the fault-tolerant behavior of the program, as reasoning may still be continued in absence of some program components, crashed because of hardware failure of a few units.

The book examined the scope of concurrent resolution of program clauses on a Petri net. The transitions in a Petri net keep track of the program clauses, where the literals in the head (body) part are mapped at its output (input) places. Places of the Petri net are shared by multiple program clauses. For instance, two program clauses having one common literal in the body of one clause and in the head of the other clause is denoted by a common place. This place is an input place of the transition describing the former clause and an output place of the transition describing the latter clause. The constant arguments of predicates present in body-less clauses are also mapped at the input place of transitions representing clauses having common literal in its body. Such organization of the Petri net framework facilitates the scope of concurrent resolution of multiple program clauses mapped onto the input/output or both types of places associated with the transitions. Further, under favorable conditions the program resources associated with more than one transition may also participate in the concurrent resolution process, thereby increasing the throughput of the system to a great extent.

The concurrent resolution introduced above automatically takes care of the AND-, OR-, Stream- and Unification-parallelisms. Consequently, the Petri net architecture provides a strong foundation to logic programming with high degree of parallelism. A suitable realization of the Petri net topology on logic architecture, therefore, is a good choice for an alternative hardwired inference engine for logic programming. The latter part of the book is, therefore, devoted to designing logic architecture for Petri net like inference engine. The proposed logic architecture comprises of six main modules, where multiplicity of the modules are needed to handle the concurrent resolution of clauses at more than one transition simultaneously.

For convenience of computational benefits, the concurrent resolution process in the proposed architecture has been realized in two elementary steps. The first step attempts to identify the possible bindings of the variable arguments of the predicates mapped at the places of the Petri net. The second step checks the consistency of the variable bindings in the arguments of the predicates mapped at the input and the output place of each transition. The first step is called the local matching (assignment, to be more specific), and the second step is referred to as global matching. The global matching helps in determining the *most general unifier* (*mgu*) of the resolved clauses, which later provides a solution to the goal/sub-goal predicate, mapped at the input/output place of each transition. After a sub-goal at a given place is determined in the manner described above, it becomes part of another clause sharing that place. Consequently, the clause thus re-organized may participate in the resolution process in a subsequent time, when its descriptor transition has a consistent set of variable bindings in the predicates mapped at its connected input/output places. The *mgu* thus obtained at one transition helps a neighboring transition to fire (resolve clauses), and the process continues until no new *mgu* at the transitions can be derived.

The architecture designed for the proposed inference engine comprises of four pipelined stages. In the first stage the places associated with each transition are activated for local token matching, whereas the second stage executes the local token matching. The third stage performs the global token matching, and the last stage ensures firing of appropriate transitions for token transfer to its associated (inert) place that did not participate in the resolution process. The approximate time required for firing a transition is around 25 T_c, where T_c denotes the time period of the system clock. The speed-up factor for the proposed inference engine for a program with n clauses and k number of concurrent set of resolvable clauses is O (n/k).

2.10 Conclusions

An examination of the existing parallel models of logic programs reveals that none of the models are sufficient to realize all four possible parallelisms in a logic program. This indicates that a specialized architecture capable of supporting all the four parallelisms is yet to be inherited. The data structure of a Petri net model in view of the above requirement is examined in this chapter. It is indeed important to note that a Petri net model, if properly designed, can be utilized to realize the above four parallelisms of a logic program. The rest of the book stresses the importance of Petri net model and proposes a new framework of extended Petri met to support the parallelisms in a logic program.

Exercises

1. Consider the following logic program containing concurrent AND-OR parallelism.

   Logic program:

   > Compute $(X, Z) \leftarrow$ Pick-up (X, Y), Square (Y, Y_2), Cube (Y, Y_3), Subtract (Y_3, Y_2, Z).
   > Square $(X, Y) \leftarrow Y = X * X$.
   > Cube $(X, Y) \leftarrow Y = X * X * X$.
   > Subtract $(X, Y, Z) \leftarrow Z = X - Y$.
   > Pick-up $([X|L], Y) \leftarrow Y = X$.
   > Pick-up $([-|L], Y) \leftarrow$ Pick-up (L, Y).

   (a) Draw a schematic dataflow graph for the above program representing both pipelining and parallelism of operations to execute the program.

   (b) Given $X = [1, 2, 3]$ and $Y = [v (1, \alpha_1), v (2, \alpha_2), v (3, \alpha_3)]$, where $\alpha_1, \alpha_2, \alpha_3$ denote three colors, using graph coloring scheme on the dataflow graph how can you compute the result as an outcome of this graph?

   [Hints:

   (a) Here, Pick-up is an OR-predicate and Pick-up, Square, Cube, Subtract are AND-predicates. If we consider the list $L = [1, 2, 3]$. When AND- and OR-parallelisms are allowed together, the parameter passing by the Pick-up OR- predicate to the evaluation process of Square and Cube is done in pipeline, and then the subtraction process becomes active. A schematic diagram of the overall system is presented in Fig. 2.12.

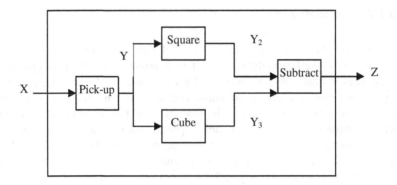

Fig. 2.12: The data flow graph of compute

(b) With X = [1, 2, 3] and Y = [v (1, α_1), v (2, α_2), v (3, α_3)],
we evaluate Y_2 and Y_3 vectors, and finally Z as follows:
$$Y_2 = [v (1, \alpha_1), v (4, \alpha_2), v (9, \alpha_3)]$$
and
$$Y_3 = [v (1, \alpha_1), v (8, \alpha_2), v (27, \alpha_3)]$$
and consequently Z takes the form
$$Z = [v (0, \alpha_1), v (4, \alpha_2), v (18, \alpha_3)]. \qquad]$$

2. For the following logic program, identify the guarded clause and also indicate
the head part, the guard part and the body part. Show the possible parallelism
in concurrent resolution of the guarded clause, if any, with others.
Logic program:

Cl_1: Find (Sum (X-square, Y-cube), Z) ←Int (X), Int (Y),
    Greater-than (Y, X)│ Find (X-square, Z_1),
    Find (Y-cube, Z_2), Sum (Z_1, Z_2, Z_3).
Cl_2: Int (1) ←.
Cl_3: Int (2) ←.
Cl_4: Int (3) ←.

[**Hints:** According to the definition 2.1, Cl_1 is a guarded clause where Find
(Sum (X-square, Y-cube), Z) is the head part; Int (X), Int (Y), Greater-than
(Y, X) is the guarded part and Find (X-square, Z_1), Find (Y-cube, Z_2), Sum
(Z_1, Z_2, Z_3) is the body part.

In the process of resolution, AND-parallelism takes place when Cl_2, Cl_3;
Cl_3, Cl_4; Cl_2, Cl_4 are resolved in parallel with Cl_1. Again, OR-parallelism
takes place when the clauses Cl_2, Cl_3, Cl_4 are attempted to resolve with Cl_1 for
Int (X) or Int (Y) in parallel.]

3. Apply rule of suspension to determine the order of computation of the given guarded clause.

    Cl_1: G ←Find (Sum (X-square, Y-cube, Z)).
    Cl_2: Find (Sum (X-square, Y-cube), Z) ←Int (X), Int (Y), Greater-than (Y, X) | Find (X-square, Z_1), Find (Y-cube, Z_2), Sum (Z_1, Z_2, Z_3).
    Cl_3: Int (1) ←.
    Cl_4: Int (2) ←.
    Cl_5: Int (3) ←.

    [**Hints:** Let us assume that there exists no other clauses except clauses Cl_3, Cl_4 and Cl_5, to instantiate the guarded part of Cl_2. Under this case Cl_2 will be instantiated first with Cl_3 and Cl_4 using AND-parallelism and next Cl_2 will be reinstantiated with Cl_4 and Cl_5 and finally with Cl_3 and Cl_5. Only after all these three sets of concurrent resolutions (AND-parallelism) the guarded clause Cl_2 can be instantiated with Cl_1. Thus the instantiation of Cl_2 with Cl_1 is withheld (suspended) for a time duration until instantiation of the guarded part of Cl_2 with all possible clauses is over.]

4. Apply the rule of commitment to justify the computation of the guarded clause.

    Cl_1: G ←Find (Sum (X-square, Y-cube), Z), Find-product (Z * 2, P).
    Cl_2: Find (Sum (X-square, Y-cube), Z) ←Int (X), Int (Y), Greater-than (Y, X) | Find (X-square, Z_1), Find (Y-cube, Z_2), Sum (Z_1, Z_2, Z_3).
    Cl_3: Int (1) ←.
    Cl_4: Int (2) ←.
    Cl_5: Int (3) ←.

    [**Hints:** Since 'Find-product (Z * 2, P)' of clause Cl_1 cannot be instantiated with any other available clauses, 'Find (Sum (X-square, Y-cube), Z)' of Cl_1 will be instantiated with the head of Cl_2. Thus Cl_2 is committed to instantiation with Cl_1.]

5. Represent the following logic program as an AND-OR tree of three levels where each level indicates the exploration of the AND/OR literals through resolution with suitable clauses.

    Logic Program:

    Cl_1: A ←a_1.
    Cl_2: A ←a_2.
    Cl_3: B ←b_1.
    Cl_4: B ←b_2.

Cl_5: C ←c_1.
Cl_6: C ←c_2.

Query: ←A, B, C.

[**Hints:** Figure 2.13 represents the AND-OR tree for the given logic program.

The first level is the AND node, the second level denotes the OR nodes and the third level denotes the AND nodes. The AND clauses at the first level are searched for exploring them in parallel. Then the clauses are resolved with the appropriate clauses.

6. The logic program given below includes both AND- and OR-parallelism. Construct an AND-OR tree using Kale's formula and show that for the following clauses unrestricted AND-/OR- parallelism can be exploited.

   Logic Program:

   Cl_1: A(a, b) ←.
   Cl_2: A(c, d) ←.
   Cl_3: B(d, e) ←.
   Cl_4: B(f, g) ←.

   Query: ←A(X, Y), B(U, V).

[**Hints:** Figure 2.14 demonstrates the AND-OR tree constructed using Kale's formula and it is evident from the Fig.2.14 that the AND nodes and the OR nodes are searched in parallel for resolution according to the Fig. 2.14.

7. The following logic program includes both AND-OR parallelism, but a search is needed to determine the consistent bindings of the variables. Using Kale's AND-OR tree show how restricted AND-OR parallelism can be handled to find a solution for the problem.

   Logic Program:

   Cl_1: A(a, b) ←.
   Cl_2: A(c, d) ←.
   Cl_3: B(b, a) ←.
   Cl_4: B(d, c) ←.
   Cl_5: B(d, e) ←.

   Query: ←A(X, Y), B(Y, X).

[**Hints:** The AND-OR tree is constructed for the given logic program vide Fig. 2.15.

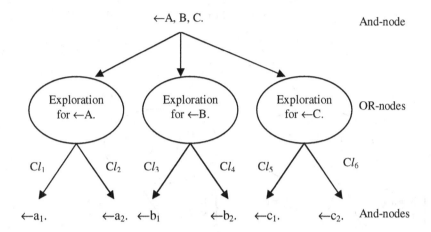

And-node

OR-nodes

Cl_1 Cl_2 Cl_3 Cl_4 Cl_5 Cl_6

And-nodes

Resulting substitution

Fig. 2.13: The AND-OR tree of three levels   ]

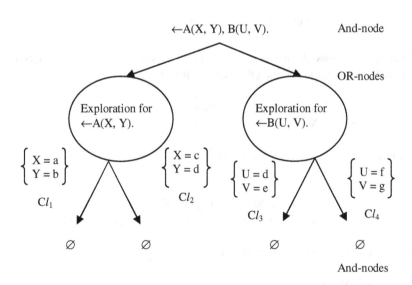

Fig. 2.14: The AND-OR tree of three levels   ]

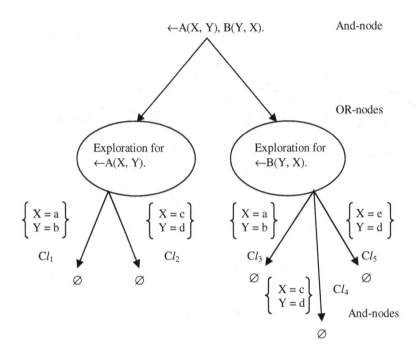

Fig. 2.15: The AND-OR tree

Let θ_1 be the possible instantiations of the OR node '$\leftarrow$A(X, Y).' and θ_2 be the possible instantiation of the OR node '$\leftarrow$B(Y, X).'. Thus,

$$\theta_1 = \{(X, Y): <a, b> \mid <c, d>\}$$
$$\theta_2 = \{(X, Y): <a, b> \mid <c, d> \mid <e, d>\}$$

Therefore, the solution set S is given by

$$S = \theta_1 \cap \theta_2$$
$$= \{(X, Y): <a, b> \mid <c, d>\} \text{ i.e., } X = a, Y = b \text{ and}$$
$$X = c, Y = d \text{ are the possible solution of the given program.]}$$

8. Given the following logic program and the query:

Logic Program:

Cl_1: Equal(X, Least-of(X, Y, Z)) $\leftarrow$X<Y, X<Z.

Query: $\leftarrow$Equal(X, Least-of(6, 4, 9)).

(a)   Construct an SLD-tree for the above logic program and show failure and backtracking to the root, and also success, when a suitable value of X is found.

(b)   Construct an AND-OR tree following Kale and show how AND- and OR-parallelism can take place in this context. Does this program include Restricted AND parallelism?

[**Hints:**

(a)   The SLD-tree for the aforementioned logic program has been constructed as shown in the Fig. 2.16.

←Equal (X, Least-of (6, 4, 9)).

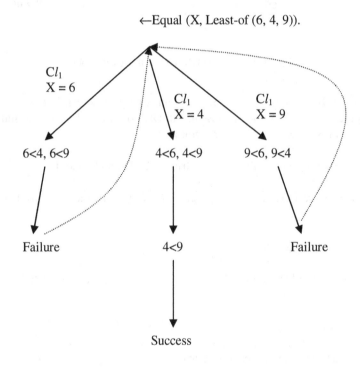

Fig. 2.16: The SLD-tree for the given problem

When X = 6 is considered, failure takes place as 6 is not less than 4. Considering X = 4, we get success as 4 is less than both 6 and 9. Again, when we consider X = 9, we get failure as 9 is not less than 4 or 6. When failure occurs, backtracking takes place to the root of the SLD-tree as shown in the Fig. 2.16.

(b) An AND–OR tree following Kale is shown in Fig. 2.17.

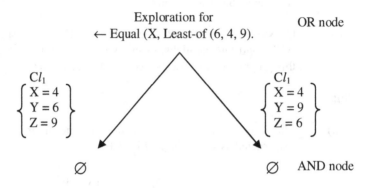

Fig. 2.17: The AND-OR tree for the given problem

An AND-OR tree following Kale is constructed as shown in Fig. 2.17. Here, as the initial AND node is absent, the OR nodes are shown determining the value of X given in the logic program.

This program does not include Restricted AND-parallelism.]

9. Consider the following logic program for testing Pythagorean triplets (X, Y, Z).

Logic Program:

Cl_1: Find Pythagorean-triplets(X, Y, Z) ←Integer(X), Integer(Y), Integer(Z), Y>X, Z>Y, $Z^2 = X^2 + Y^2$.

Query: ←Find Pythagorean-triplets (X, Y, Z).

(a) List the order of generating and testing the values of (X, Y, Z) to satisfy Pythagorean-triplets (X, Y, Z).

(b) Construct a Data Join Graph (DJG) to represent the order of generating and testing (X, Y, Z) for the given problem.

(c) Verify whether the following goal instances satisfy the Pythagorean-triplets (X, Y, Z) on the DJG.

Pythagorean-triplets $(1, 2, 3) \leftarrow$.
Pythagorean-triplets $(3, 4, 5) \leftarrow$.
Pythagorean-triplets $(6, 8, 10) \leftarrow$.

[**Hints:**

(a) The order of the generation and testing of (X, Y, Z) are presented in consecutive lines.

Generate an integer X.
Generate an integer Y>X.
Generate an integer Z>Y.
Test whether $Z^2 = X^2 + Y^2$.

(b) A Data Join Graph is constructed (vide Fig. 2.18) to represent the order of generating and testing (X, Y, Z) for the given problem.

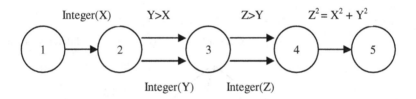

Fig. 2.18: Data Join Graph representing the order of generating and testing (X, Y, Z) for the given problem

(c) Here we verify with the second instance only (vide Fig. 2.19).

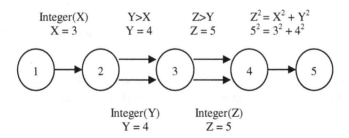

Fig. 2.19: Data Join Graph verifying the second instance for the given problem

The reader can verify the other two instances himself/herself.]

10. (a) Construct an SLD-tree for the following logic program and mark the inference depth of the nodes in the tree.

Logic Program:

Cl_1: Equal(X, Sum(1, 2)) ←.
Cl_2: Equal(X, Sum(3, 2)) ←.

Query: ←Equal(X, Sum(A, B), A>B).

(b) Also construct a CAM with appropriate fields to describe backtracking on the SLD-tree.

[Hints:

(a) The SLD-tree for the given logic program is given in the Fig. 2.20. The inference nodes are also marked in the Fig. 2.20.

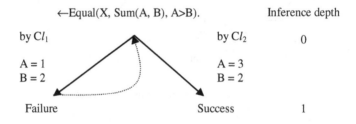

Fig. 2.20: The SLD-tree for the given logic program

(b) A CAM is constructed as shown in the Fig. 2.21 for the given problem.

The garbage collected in the second step is removed in the third step. In the fourth step, a new set of value for A and B is considered.

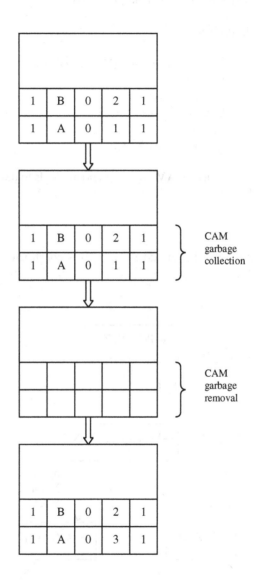

Fig. 2.21: A CAM for the given problem   ]

11. (a)   Construct an SLD-tree for the following logic program and mark the
           inference depth of the nodes in the tree.

Logic Program:

Cl_1: P(X, Y, Z) ←W(Y, X), R(Z, X).
Cl_2: W(d, e) ←.
Cl_3: W(b, a) ←.
Cl_4: R(c, a) ←.

Query: ←P(X, Y, Z).

(b) Also construct a CAM with appropriate fields to describe backtracking on
     the SLD-tree.

[Hints:

(a) The SLD-tree for the given problem is constructed as shown in the Fig.
    2.22.

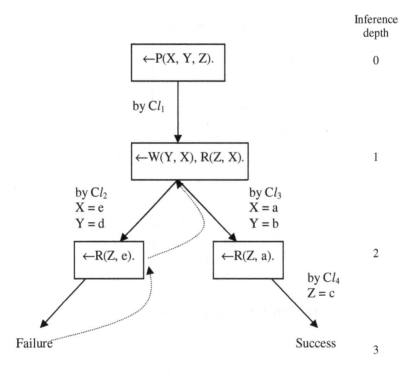

Fig. 2.22: The SLD-tree for the given problem

When Cl_2 was considered for resolution, no clause is available for resolving with the resolvent. Therefore, failure takes place. But when Cl_3 is considered, Cl_4 readily resolves with the resolvent and 'success' is obtained at the level three as shown in the Fig. 2.22.

(b) A CAM is constructed as shown in the Fig. 2.23 for the given problem.

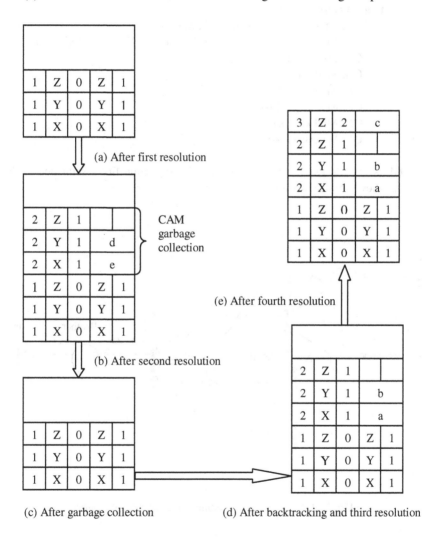

(a) After first resolution

CAM garbage collection

(b) After second resolution

(c) After garbage collection

(d) After backtracking and third resolution

(e) After fourth resolution

Fig. 2.23: The CAM demonstrating the CAM garbage removal]

12. (a) Represent the following logic program by the Petri Net model of Murata.

   (b) Construct the incidence matrix A for the network of part (a), and hence
        determine the time invariant solution X, where A∘X = 0.

   Logic Program:
   Cl_1: Father(X, Y) ←Son(Y, X), Male(X).
   Cl_2: Brother(Y, Z) ←Father(X, Y), Mother(W, Z), Wife(W, X).
   Cl_3: Son(l, r) ←.
   Cl_4: Son(k, r) ←.
   Cl_5: Male(r) ←.
   Cl_6: Mother(s, l) ←.
   Cl_7: Mother(s, k) ←.
   Cl_8: Wife(s, r) ←.

[**Hints:** Two possible sets of clauses following same firing sequence of the
transitions are listed below:

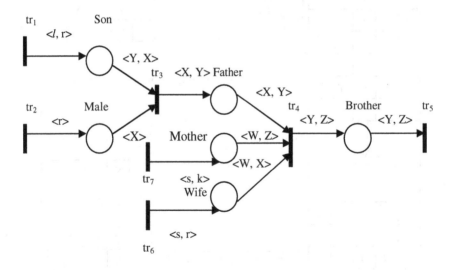

Fig. 2.24: Petri net corresponding to the given logic program

Sequence: $tr_1/tr_2 - tr_3 - tr_6/tr_7 - tr_4 - tr_5$.

   A Petri net is drawn for the given logic program considering the clauses
Cl_1, Cl_2, Cl_3, Cl_5, Cl_7, Cl_8 vide Fig. 2.24.

Here, as the generic state equation of a logic program realized with a Petri net is given by

$$A \circ X = 0$$

where $A = [a_{ij}]$ is an incidence matrix, X provides a solution to the user's query and o denotes a matrix product with substitution. The following incidence matrix is worked out from the said logic program also.

	Son	Male	Father	Mother	Wife	Brother
tr_1	$<l, r>$	0	0	0	0	0
tr_2	0	$<r>$	0	0	0	0
tr_3	$-<Y, X>$	$-<X>$	$<X, Y>$	0	0	0
$A = tr_4$	0	0	$-<X, Y>$	$-<W, Z>$	$-<W, X>$	$<Y, Z>$
tr_5	0	0	0	0	0	$-<Y, Z>$
tr_6	0	0	0	0	$<s, r>$	0
tr_7	0	0	0	$<s, k>$	0	0

Now, A^T, the transpose of A can be written as follows:

	tr_1	tr_2	tr_3	tr_4	tr_5	tr_6	tr_7
Son	$<l, r>$	0	$-<Y, X>$	0	0	0	0
Male	0	$<r>$	$-<X>$	0	0	0	0
Father	0	0	$<X, Y>$	$-<X, Y>$	0	0	0
A^T=Mother	0	0	0	$-<W, Z>$	0	0	$<s, k>$
Wife	0	0	0	$-<W, X>$	0	$<s, r>$	0
Brother	0	0	0	$<Y, Z>$	$-<Y, Z>$	0	0

The solution vector X_1 for the above firing sequence is computed as follows:

$$X_1 = \begin{array}{c} tr_1 \\ tr_2 \\ tr_3 \\ tr_4 \\ tr_5 \\ tr_6 \\ tr_7 \end{array} \left(\begin{array}{c} \varnothing \\ \{\,\} \\ \{l/Y, r/X\} \\ \{r/X, l/Y, k/Z, s/W\} \\ \{\,\} \\ \{s/W, r/X\} \\ \{\,\} \end{array} \right)$$

A Petri net is drawn for the given logic program considering the clauses Cl_1, Cl_2, Cl_4, Cl_5, Cl_6, Cl_8 vide Fig. 2.25 for obtaining the second solution X_2

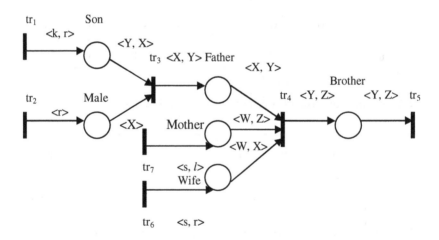

Fig. 2.25: Petri net corresponding to the given logic program

Here, once again, as the generic state equation of a logic program realized with a Petri net is given by

$$A \circ X = 0$$

where $A = [a_{ij}]$ is an incidence matrix, X provides a solution to the user's query and o denotes a matrix product with substitution, the following incidence matrix is worked out from the said logic program also.

	Son	Male	Father	Mother	Wife	Brother
tr_1	<k, r>	0	0	0	0	0
tr_2	0	<r>	0	0	0	0
tr_3	-<Y, X>	-<X>	<X, Y>	0	0	0
tr_4	0	0	-<X Y>	-<W, Z>	-<W, X>	<Y, Z>
tr_5	0	0	0	0	0	-<Y, Z>
tr_6	0	0	0	0	<s, r>	0
tr_7	0	0	0	<s, l>	0	0

$A =$ (matrix above)

Now, A^T, the transpose of A can be written as follows:

	tr_1	tr_2	tr_3	tr_4	tr_5	tr_6	tr_7
Son	<k, r>	0	-<Y, X>	0	0	0	0
Male	0	<r>	-<X>	0	0	0	0
Father	0	0	<X Y>	-<X, Y>	0	0	0
Mother	0	0	0	-<W, Z>	0	0	<s, l>
Wife	0	0	0	-<W, X>	0	<s, r>	0
Brother	0	0	0	<Y, Z>	-<Y, Z>	0	0

$A^T =$ (matrix above)

Now, the solution X_2 is given by,

$$X_2 = \begin{array}{c} tr_1 \\ tr_2 \\ tr_3 \\ tr_4 \\ tr_5 \\ tr_6 \\ tr_7 \end{array} \left(\begin{array}{c} \varnothing \\ \{\ \} \\ \{k/Y, r/X\} \\ \{r/X, k/Y, l/Z, s/W\} \\ \{\ \} \\ \{s/W, r/X\} \\ \{\ \} \end{array} \right)$$

It was indeed important to note that the above two solutions are time invariant and therefore were generated by satisfying the equations $A^T o X_1 = 0$ and $A^T o X_2 = 0$.

13. Given a **logic program:**

Cl_1: Brother(Y, Z) ←Father(X, Y), Mother(W, Z), Wife(W, X).
Cl_2: Mother(s, l) ←.
Cl_3: Mother(s, k) ←.
Cl_4: Wife(s, r) ←.
Cl_5: Father(r, l) ←.
Cl_6: Father(r, k) ←.

(a) Using the following firing sequence of the transitions, determine the answer of the query:

Brother(l, Z) ←.

Firing order of transitions: $tr_1 — tr_5$

(b) Is this firing order unique?

[Hints:

(a) With the given query as the token in the place indicated by the predicate, transition tr_1 fires and a new token is generated in the place 'Mother'. As the token can be matched with the arc function $<s, k>$, tr_5 can be fired. Then the Petri net will have an empty marking indicating a successful sequence of firing (vide Fig. 2.26).

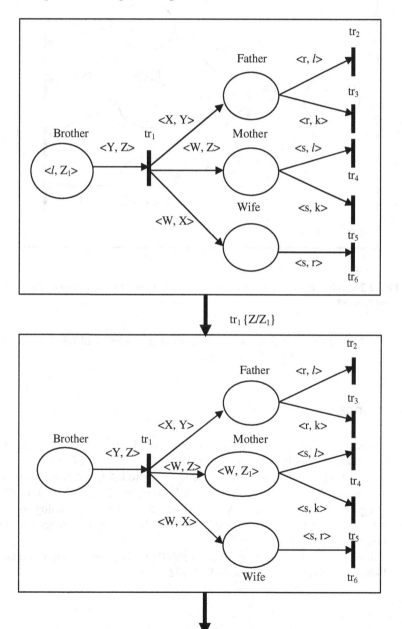

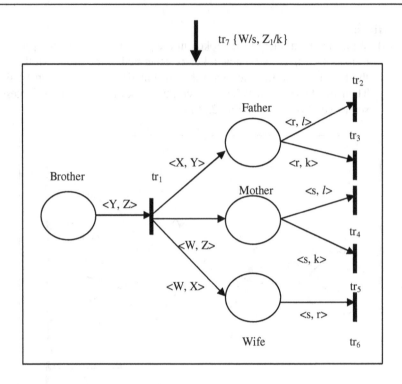

Fig. 2.26: The Petri net representation for the successful firing sequence of the given logic program

(b) No, because on firing of the transition tr_1 traversal of token to no other place except 'Mother' is feasible.]

References

1. Bender, E. A., *Mathematical Methods in Artificial Intelligence*, IEEE Computer Society Press, Los Alamitos, CA, chapter 1, pp. 26, 1996.
2. Chang, J. H. and Despain, A., "Extending a PROLOG machine for parallel execution," *Proc. 1978 of Int. Symp. on Logic Prog.*, Boston, MA, July 1985.
3. Ganguly, S., Silberschatz, A. and Tsur, S., "Mapping datalog program execution to networks of processors," *IEEE Trans. on Knowledge and Data Engg.*, vol. 7, no. 3, June 1995.
4. Halder, S., *On the Design of Efficient PROLOG Machines*, M.E. dissertation, Jadavpur University, Calcutta, India, 1992.

5. Hermenegildo, M. and Tick, E., "Memory performance of AND-parallel PROLOG on shared-memory architecture," *Proc. of the 1988 Int. Conf. on Parallel Processing*, vol. II, Software, pp. 17-21, Aug. 15-19, 1988.

6. Jeffrey, J., Lobo, J., Murata, T., "A high-level Petri net for goal-directed semantics of Horn clause logic," *IEEE Trans. on Knowledge and Data Engg.*, vol. 8, no. 2, April 1996.

7. Kale, M. V., "Parallel problem solving," in *Parallel Algorithms for Machine Intelligence and Vision*, Kumar, V., Gopalakrishnan, P. S. and Kanal, L. N., (Eds.), Springer-Verlag, Heidelberg, 1990.

8. Konar, A. and Mandal, A. K., "Uncertainty management in expert systems using fuzzy Petri nets," *IEEE Trans. on Knowledge and Data Engg.*, vol.8, no. 1, Feb. 1996.

9. Li, L., "High-level Petri net model of logic program with negation," *IEEE Trans. on Knowledge and Data Engg.*, vol. 6, no. 3, June 1994.

10. Murata, T., "Petri nets: Properties, Analysis and Applications," *IEEE Proc.*, vol. 77, no. 4, April 1989.

11. Murata, T. and Yamaguchi, H., "A Petri net with negative tokens and its application to automated reasoning," *Proc. of the 33rd Midwest Symp. on Circuits and Systems*, Calgary, Canada, Aug. 12-15, 1990.

12. Naganuma, J., Ogura, T., Yamada, S-I., Kimura, T., "High-speed CAM-based architectutre for a PROLOG machine (ASCA)," *IEEE Trans. on Computers*, vol. 37, no. 11, Nov. 1988.

13. Patt, Y. N., "Alternative implementations of PROLOG: the micro architecture perspectives," *IEEE Trans. on Systems, Man and Cybernetics*, vol. 19, no. 4, July/ Aug. 1989.

14. Peterka, G. and Murata, T., "Proof procedure and answer extraction in Petri net model of logic programs," *IEEE Trans. on Software Engg.*, vol. 15, no. 2, Feb. 1989.

15. Takeuchi, A., *Parallel Logic Programming*, Wiley, NY, 1992.

16. Takeuchi, A., "On an extension of stream-based AND-parallel logic programming languages," in *proc. of the First National Conf. of Japan Society for Software Science and Technology*, pp. 291-294, 1984.

17. Takeuchi, A., Takahashi, K. and Shimizu, H., A parallel problem solving language for concurrent systems, in *Concepts and Characteristics of Knowledge-based Systems*, Tokoro, M., Anzai, Y. and Yonezawa, A., (Eds.), Elsevier, North Holland, pp. 267-296, 1989.

18. Takeuchi, A. and Takahashi, K., An operational semantics of AND-OR II: A parallel logic programming language with AND- and OR- parallelism, in *Concurrency: Theory, Language and Architecture*, UK/Japan Workshop, Lecture Notes in Computer Science, Springer-Verlag, vol. 491, Oxford, UK, Sept. 1989.

19. Yan, J. C., "Towards parallel knowledge processing," in *Advanced series on Artificial Intelligence, Knowledge Engineering Shells: Systems and Techniques*, Bourbakis (Ed.), vol. 2, World Scientific, Singapore, 1993.

3

The Petri Net Model — A New Approach

The chapter presents a new approach to reason with logic programs using a specialized data structure similar to Petri nets. Typical logic programs include a number of concurrently resolvable clauses. A priori detection of these clauses indeed is useful for their subsequent participation in the concurrent resolution process. The chapter explores the scope of distributed mapping of program clause components onto Petri nets so as to automatically select the participant clauses for concurrent resolution. An algorithm for concurrent resolution of clauses on Petri nets has been undertaken with a motive to improve the speed-up factor of execution of the program without sacrificing the resource utilization rate. All the new concepts have been illustrated with examples. The exercise at the end of the chapter includes a number of interesting problems provided with sufficient hints to enable the readers to verify their understanding.

3.1 Introduction

Logic programming has already gained much importance for its increasing applications in data and knowledge engineering. A logic program usually consists of a special type of program clauses known as ***Horn clauses***. Programs built with Horn clauses only are called normal logic programs. Complex knowledge having multiple consequent literals cannot be represented by normal logic programs because of its structural restriction imposed by Horn clauses.

In spite of its limitations in knowledge representation, normal logic programs are still prevalent in relational languages like PROLOG and designing DATALOG for simplicity in their compilers. Generally, compilers for logic programming employ SLD-resolution that resolves each pair of program clauses at a time. Execution of a program by SLD-resolution thus requires a considerable amount of time. Most logic programs usually include a number of concurrently resolvable clauses; unfortunately there is hardly any literature on parallel and distributed models of logic programming, capable of resolving multiple program clauses concurrently.

In early nineties, Patt [13] examined a non-conventional execution model of a uniprocessor micro-engine for a PROLOG program and measured its performance

A. Bhattacharya et al.: *The Petri Net Model — A New Approach*, Studies in Computational Intelligence
(SCI) **24**, 107–175 (2006)
www.springerlink.com   © Springer-Verlag Berlin Heidelberg 2006

with a set of 14 benchmarks. Yan [16] provided a novel scheme for concurrent realization of PROLOG on a RAP-WAM machine. The WAM generates an intermediate code during the compilation phase of a PROLOG program that exploits the fullest degree of parallelism in the program itself. The RAP, on the other hand, reduces the overhead associated with the run-time management of *variable binding conflict* between goals.

Hermenegildo and Tick proposed an alternative model [4] for concurrent execution of PROLOG programs by RAP-WAM combinations by representing the dependency relationship of the program clauses by a graph, where the take-off arcs at the vertex in the graph denotes parallel tasks in the program. They employed goal stacks with each processing element. When a parallel call is invoked, the concurrent tasks are mapped autonomously to the stacks of the less busy or idle processing elements. A synchronization and co-ordination for parallel calls was also implemented in their scheme.

Recently Ganguly, Silberschatz and Tsur [3] presented a new algorithm for automated mapping of logic programs onto a parallel processing architecture. In their first scheme, they considered the mapping of program clauses having shared variables onto adjacent processors. This reduces the communication overhead among the program clauses. In the latter part of their work, they eliminated the above constraints at the cost of extra network management time.

Takeuchi in his recent book [15] presented a new language for AND-OR parallelism. Kale in a book chapter [7] discussed the scope of an alternative formulation of problem-solving using parallel AND-OR trees. Among the existing speed-up schemes of logic programming machines, the content addressable memory (CAM)–based architecture of PROLOG machines by Naganuma et al. [12] needs special mention. To speed up the execution performance of PROLOG programs, they employed *hierarchical pipelining and garbage collection mechanism* of a CAM for efficient backtracking on a SLD tree.

Though a number of techniques are prevalent for the realization of logic programs on a parallel architecture, none of these are capable of representing the theoretically possible maximum parallelism in a program. For realization of all types of parallelism in a logic program, a specialized data structure appropriate for representing the possible parallelism is needed. Petri net has already proved itself as a successful data structure for reasoning with complex rules. For instance, rules having more than one antecedent and consequent clause with each clause containing a number of variables can easily be represented by a Petri net structure [9]. Murata [11] proposed the scope of Petri net models for knowledge representation and reasoning under the framework of predicate logic.

There is also a good many literature [6, 10, 14] dealing with much more complex reasoning problems using Petri nets. Because of the distributed structure of a Petri net, pieces of knowledge fragmented into components can easily be mapped onto this structure [8]. For example, a transition firing in a Petri net may synonymously be used as rule firing in an expert system. The transitions may be regarded as the implication operator of a rule, while its input and output places may respectively be regarded as the antecedent and consequent clauses of a rule [1]. The arguments associated with the predicates of a clause are also assigned at the arcs connecting the place containing the predicate and the transition describing the implication rule. Such fragmentation and mapping of the program components onto different modules of a Petri net enhances the scope of parallelism in a logic program. The objective of the chapter is to fragment a given logic program to smallest possible units, and map them onto a Petri net to fully exploit its parallelism.

The methodology of reasoning presented in the chapter is an extension of Murata's classical models on Petri nets [11, 14], applied to logic programming. Murata defined a set of rules to synonymously describe the resolution of horn clauses in a normal logic program with the firing of transitions in a Petri net. The chapter attempts to extend Murata's scheme for automated reasoning to non-Horn clause based programs as well.

Section 3.2 provides related definitions of important terminologies used in this chapter. The concept of concurrency in resolution process is introduced in section 3.3. A new model for concurrent resolution on Petri nets is presented in section 3.4. An algorithm for concurrent resolution is presented in section 3.5. Performance analysis of the Petri net-based model is covered in section 3.6. Conclusions are listed in section 3.7. A set of numerical problems has been undertaken in the exercise of the chapter.

3.2 Formal Definitions

In this section, we provide relevant definitions to logic programming and different methodologies to execute logic programs by resolution of program clauses.

3.2.1 Preliminary Definitions

Definition 3.1: *A **clause** cl_i is represented by*

$$A_i \leftarrow B_i . \tag{3.1}$$

where B_i denotes the body, A_i denotes the head and '$\leftarrow$' denotes the implication operator.

The body B_i usually is a conjunction of literals $B_{ij} \, \exists j$, i.e.,

$$B_i = B_{i1} \wedge B_{i2} \wedge \ldots \wedge B_{ij} \tag{3.2}$$

The head A_i usually is a disjunction of literals $A_{ik} \, \exists k$, i.e.,

$$A_i = A_{i1} \vee A_{i2} \vee \ldots \vee A_{ik} \tag{3.3}$$

The literals A_{ik} and B_{ij} have arguments containing terms that may include variables (denoted by capital letters), constants (denoted by small letters), function of variables and function of function of variables (in a recursive form).

Example 3.1: This example illustrates the constituents of a clause. For instance,

$$\text{Father(Y, Z), Uncle(Y, Z)} \leftarrow \text{Father(X, Y), Grandfather(X, Z).} \quad (3.4)$$

is an example of a general clause, where the body consists of Father (X, Y) and Grandfather (X, Z) and the head consists of Father (Y, Z) and Uncle (Y, Z). The clause states that if X is the father of Y and X is the grandfather of Z, then either Y is the father of Z or Y is the uncle of Z.

In case all the terms are bound variables or constants, the literals A_{ik} or B_{ij} are called **ground literals**.

Example 3.2: The following is an example of a clause with all variables been bound by constants, thereby resulting in ground literals: Father (n, a) and Son (a, n).

$$\text{Father(n, a)} \leftarrow \text{Son(a, n).} \quad (3.5)$$

The above clause states that *if a is son of n then n is the father of a.*

Special cases:

(i)   In case of a **goal clause (query)** the consequent part A_i is absent.

The clause (3.6) presented below contains no consequent part, and hence it is a query.

$$\leftarrow \text{Grandfather(X, Z).} \quad (3.6)$$

Given that X is the grandfather of Z, the clause (3.6) questions the value of X and Z.

(ii) A clause with an empty body and consisting of ground literals in the head is regarded as a *fact.*

The clause (3.7) below contains no body part and the variable arguments are bound by constants. Thus it is a fact.

$$\text{Grandfather(r, a)} \leftarrow. \quad (3.7)$$

It states that r is grandfather of a.

(iii) When the consequent part A_i includes a single literal, the resulting clause is called a ***Horn clause***. The details about Horn clause are given in definition 3.2.

(iv) When A_{ik} and B_{ij} do not include arguments, we call them propositions and the clause '$A_i \leftarrow B_i$.' is then called a ***propositional clause***.

The clause (3.8) below for instance is a propositional clause as it does not contain any arguments.

$$P \leftarrow Q, R. \tag{3.8}$$

Definition 3.2: *A **Horn clause** contains a head and a body with at most one literal in its head.*

Example 3.3: The clause (3.9) is an example of a Horn clause.

$$P \leftarrow Q_1, Q_2, \ldots, Q_n. \tag{3.9}$$

It represents a Horn clause where P and the Q_i are literals or atomic formulas. It means if all the Q_is are true, then P is also true. Q_i is the body part and P is the head in this Horn clause.

Definition 3.3: *The clauses containing more than one literal in its head are known as **Non-Horn Clauses**.*

Example 3.4: The clause (3.10) for instance is a non-Horn clause.

$$P_1, P_2, \ldots, P_m \leftarrow Q_1, Q_2, \ldots, Q_n. \tag{3.10}$$

Definition 3.4: *An **Extended Horn Clause (EHC)** contains a head and a body with at least one clause in the body and zero or more number of clause in its head. Commas are used to denote conjunction of the literals in the body and disjunction of literals in the head.*

Example 3.5: The general format of an EHC is

$$A_1, A_2, \ldots, A_n \leftarrow B_1, B_2, \ldots, B_m. \tag{3.11}$$

Here the head and the body contain n and m number of literals respectively.

It is important to note that an extended Horn clause includes both Horn clause and its extension as well.

Definition 3.5: *A program that contains extended Horn clauses, as defined above, is called an **Extended Logic Program**.*

Example 3.6: The clauses (3.12 – 3.15) together represent an extended logic program. It includes extended Horn clause (3.12) with facts (3.13 – 3.15).

$$\text{Father(Y, Z), Uncle(Y, Z)} \leftarrow \text{Father(X, Y), Grandfather(X, Z).} \qquad (3.12)$$
$$\text{Father(r, d)} \leftarrow . \qquad (3.13)$$
$$\neg\text{Father(d, a)} \leftarrow . \qquad (3.14)$$
$$\text{Grandfather(r, a)} \leftarrow . \qquad (3.15)$$

To represent the query "whether d is uncle of a?" the goal clause of the following form may be constructed.

$$\text{Goal:} \leftarrow \text{Uncle(d, a).} \qquad (3.16)$$

The answer to the query can be obtained by taking negation of the goal and then resolving it with the supplied clauses (3.12 – 3.15). In the present context, the answer to the query will be *true*.

To explain this, we need to introduce the principles of ***resolvability of two clauses.*** In order to explain resolvability of clauses we further need to introduce ***substitution sets*** and ***most general unifier***.

Definition 3.6: *A substitution represented by a set of ordered pairs $s\{t_1/v_1, t_2/v_2,....., t_n/v_n\}$, is called the **substitution set**. The pair t_i/v_i means that the term t_i is substituted for every occurrence of the variable v_i throughout.*

Example 3.7: There exists four substitution sets for the predicate $P(a, Y, f(Z))$ in the following instances:

$$P(a, X, f(W))$$
$$P(a, Y, f(b))$$
$$P(a, g(X), f(b))$$
$$P(a, c, f(b))$$

Substitution sets for the aforementioned examples are

$$s_1 = \{X/Y, W/Z\}$$
$$s_2 = \{b/Z\}$$
$$s_3 = \{g(X)/Y, b/Z\}$$
$$s_4 = \{c/Y, b/Z\}$$

To denote a substitution instance of an expression w, using a substitution s we write ws.

Example 3.8: Let the expression w = P(a, Y, f(Z)) and the substitution set s = {X/Y, W/Z}. Then the substitution instance ws = P(a, X, f(W)).

3.2.2 Properties of the Substitution Set

Property 1: $(ws_1)\Delta s_2 = w(s_1 \Delta s_2)$ *where w is an expression and* $s_1 \Delta s_2$ *are two substitutions.*

Example 3.9: To illustrate the property 1,

let,
$$w = P(X, Y),$$
$$s_1 = \{f(Y)/X\}$$
and $s_2 = \{a/Y\}$.

Now, $(ws_1)\Delta s_2 = (P(f(Y), Y))\{a/Y\}$
$$= (P(f(a), a)).$$

Again, $w(s_1\Delta s_2) = (P(X, Y))\{f(a)/X, a/Y\}$
$$= P(f(a), a).$$

Therefore, $(ws_1)\Delta s_2 = w(s_1\Delta s_2)$.

The composition of two substitutions s_1 and s_2 in order is denoted by $s_1\Delta s_2$, which is the substitution obtained by first applying s_2 to the terms of s_1 and then adding the ordered pairs from s_2 not occurring in s_1. Example 3.10 illustrates the said concept.

Example 3.10: Let $s_1 = \{f(X, Y)/Z\}$ and $s_2 = \{a/X, b/Y, c/W, d/Z\}$.

Then $s_1\Delta s_2 = \{f(a, b)/Z, a/X, b/Y, c/W\}$.
Property 2: *Composition of substitutions is associative i.e.,*

$$(s_1\Delta s_2)\Delta s_3 = s_1\Delta(s_2\Delta s_3).$$

Example 3.11: To illustrate the associative property of composition in substitutions,
let,

$s_1 = \{f(Y)/X\}$
$s_2 = \{a/Y\}$
$s_3 = \{c/Z\}$
and $w = P(X, Y, Z)$.

Here, $(s_1\Delta s_2) = \{f(a)/X, a/Y\}$
$\quad (s_1\Delta s_2)\Delta s_3 = \{f(a)/X, a/Y, c/Z\}$

Again, $(s_2\Delta s_3) = \{a/Y, c/Z\}$
$\quad s_1\Delta(s_2\Delta s_3) = \{f(a)/X, a/y, c/Z\}$

Therefore, $(s_1\Delta s_2)\Delta s_3 = s_1\Delta(s_2\Delta s_3)$.

Property 3: *Commutability fails in case of the substitutions i.e.,*

$$s_1\Delta s_2 \neq s_2\Delta s_1.$$

Example 3.12: We can illustrate the third property following the substitution sets used in example 3.11.

Here,

$(s_1\Delta s_2) = \{f(a)/X, a/Y\}$ and $(s_2\Delta s_1) = \{a/Y, f(Y)/X\}$
So, $s_1\Delta s_2 \neq s_2 \Delta s_1$.

Definition 3.7: *Given two predicates $P(t_1, t_2, \ldots, t_n)$ and $P(r_1, r_2, \ldots, r_n)$ and $s = \{t_i/r_i\}$ is a substitution set, which on substitution in the predicates makes them identical (unifies them). Then the substitution set s is called a **unifier**. The **most general unifier** (mgu) is the simplest unifier g, so that any other unifier g′ satisfies $g′ = g \Delta s′$ for some substitution s′.*

Example 3.13: For the predicates $P(X, f(Y), b)$ and $P(X, f(b), b)$, $g′ = \{a/X, b/Y\}$ definitely is a unifier as it unifies the predicates to $P(a, f(b), b)$, but the mgu in this case is $g = \{b/Y\}$. If is to be noted further that a substitution $s′ = \{a/X\}$ satisfies $g′ = g \Delta s′$.

Definition 3.8: Let cl_i and cl_j be two clauses of the following form:

$$Cl_i \equiv A_{i1} \lor A_{i2} \lor \ldots \lor A_{ix} \lor \ldots \lor A_{im} \leftarrow B_{i1} \land B_{i2} \land \ldots \land B_{il} \tag{3.17}$$
$$Cl_j \equiv A_{j1} \lor A_{j2} \lor \ldots \lor A_{jm}{}' \leftarrow B_{j1} \land B_{j2} \land \ldots \land B_{jy} \land \ldots \land B_{jl}{}' \tag{3.18}$$

and P be the common literal present in the head of cl_i and body of cl_j. For instance, let for a substitution s

$$[A_{ix}]_s = [B_{jy}]_s = [P]_s$$

The *resolvent* of cl_i and cl_j, denoted by $R(cl_i, cl_j) = cl_{ij}$ (say), is computed as follows under the substitution $s = s_{ij}$, say.

$Cl_{ij} =$
$[(A_{i1} \lor A_{i2} \lor .....\lor A_{i(x-1)} \lor A_{i(x+1)} \lor .....\lor A_{im}) \lor (A_{j1} \lor A_{j2} \lor .....\lor A_{jm}') \leftarrow$
$(B_{i1} \land B_{i2} \land .....\land B_{il}) \land (B_{j1} \land B_{j2} \land .....\land B_{j(y-1)} \land B_{j(y+1)} \land .....\land B_{jl}')]$ (3.19)

If cl_{ij} can be computed from the given cl_i and cl_j, we say that the clauses cl_i and cl_j are *resolvable*.

Example 3.14: The clauses given are as follows:

Cl_i: Fly(X) ←Bird(X). (3.20)
Cl_j: Bird(parrot) ←. (3.21)
Cl_{ij}: Fly(parrot) ←. where $s_{ij} = \{parrot/X\}$ (3.22)

The clauses cl_i and cl_j in this example, are resolvable with the substitution s_{ij} yielding the resolvent cl_{ij}.

Definition 3.9: *If in a set of clauses there is at least one clause cl_j for each clause cl_i such that resolution holds on cl_i and cl_j, producing a resolvent cl_{ij}, then the set is called the **set of resolvable clauses**.*

Example 3.15: In the following set of clauses each clause is resolvable with at least one other clause:

Mother(Y, Z) ←Father(X, Z), Wife(Y, X). (3.23)
Father(r, l) ←. (3.24)
Wife(s, r) ←. (3.25)

Here, clause number 3.23 is resolvable with clause number 3.24 and clause number 3.25. It is an example of the set of resolvable clauses.

Now, we briefly underline Select Linear Definite (SLD)-resolution.

3.2.3 SLD Resolution

To understand SLD resolution we first have to learn a few definitions.

Definition 3.10: *A **definite program clause** is a clause of the form*

$$A \leftarrow B_1, B_2, ..., B_n.,$$

which contains precisely one atom (viz. A) in its consequent (head) and may contain a null, one or more literals in its body (viz. B_1 or B_2 or .....or B_n).

Definition 3.11: *A **definite program** is a finite set of definite program clauses.*

Definition 3.12: *A **definite goal** is a clause of the form*

$$\leftarrow B_1, B_2, ..., B_n.$$

i.e., a clause with an empty consequent.

Definition 3.13: ***SLD resolution*** *stands for **SL resolution for definite clauses**, where SL stands for **resolution with linear selection function**.*

Example 3.16: This example illustrates the linear resolution. Here, the following OR clauses (clauses connected by OR operator), represented by a set-like notation are considered.

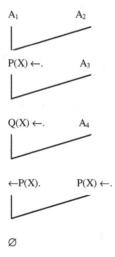

Fig. 3.1: The linear selection of clauses in the resolution tree

Let $S = \{A_1, A_2, A_3, A_4\}$,
where $A_1 = P(X), Q(X) \leftarrow$.
$\qquad A_2 = P(X) \leftarrow Q(X)$.
$\qquad A_3 = Q(X) \leftarrow P(X)$.
$\qquad A_4 = \leftarrow P(X), Q(X)$.

and goal $= \leftarrow \neg P(X)$.

By linear selection, a resolution tree can be constructed as shown in Fig. 3.1.

It is clear that two clauses from the set $S_1 = S \cup \{\neg Goal\}$ are first used for resolution with a third clause from the same set S_1. The process is continued until a null clause is generated. In the linear selection process, one clause, however, can be used more than once for resolution.

Definition 3.14: *Let $S = \{cl_1, cl_2, ..., cl_n\}$ be a set of resolvable clauses, and there exists one or more definite orders to select the clauses pair-wise for SLD resolution, without which the SLD resolution of all the n clauses fail to generate a resolvent. Under this circumstance, we say that an **orderly resolution** exists among the clauses in S.*

The resolvent of clauses cl_i and cl_j is hereafter denoted by cl_{ij} or $R(cl_i, cl_j)$, where R represents a binary resolution operator. The *order of resolution* in R ($R (cl_i, cl_j), cl_k)$ is denoted by i-j-k for brevity. It may be noted that i-j $\equiv$ j-i and i-j-k $\equiv$ k-i-j $\equiv$ k-j-i.

Example 3.17: Given the following clauses cl_1 through cl_3, we would like to illustrate the principle of orderly resolution with these clauses.

Cl_1: Paternal-uncle(X, Y), Maternal-uncle(X, Y) $\leftarrow$ Uncle(X, Y).
Cl_2: $\neg$ Paternal-uncle(n, a) $\leftarrow$. $\equiv$ $\leftarrow$ Paternal-uncle(n, a).
Cl_3: $\neg$ Maternal-uncle(n, a) $\leftarrow$. $\equiv$ $\leftarrow$ Maternal-uncle(n, a).

We now demonstrate two different orders of resolution, and show that the result is unique in both the cases. One of the orders of resolution could be 1-2-3. This is computed as follows:

$R(cl_1, cl_2)$: Maternal-uncle(n, a) $\leftarrow$ Uncle(n, a). with $s_{12} = \{n/X, a/Y\}$
$R(R(cl_1, cl_2), cl_3)$: $\leftarrow$ Uncle(n, a). $\equiv$ $\neg$ Uncle(n, a) $\leftarrow$.

An alternative order of resolution is 3-1-2. To compute this, we proceed as follows:

$R(cl_3, cl_1)$: Paternal-uncle(n, a) $\leftarrow$ Uncle(n, a). with $s_{31} = \{n/X, a/Y\}$
$R(R(cl_3, cl_1), cl_2)$: $\leftarrow$ Uncle(n, a). $\equiv$ $\neg$ Uncle(n, a) $\leftarrow$.

So, both the orderly resolutions return the same resolvent: [¬Uncle(n, a) ←.] which means n is not the uncle of a.

It is important to note that resolution of multiple clauses by different orders always do not return unique resolvent.

Definition 3.15: *If there exists only one definite order of resolution among a set of resolvable clauses, it is said to have* **single sequence** *of resolution. On occasions, we can obtain the same resolvent by taking a reversed order of resolution. For instance $R(...(R(R(R(cl_1, cl_2), cl_3), cl_4).....cl_{n-1}), cl_n) = R(...(R(R(cl_n, cl_{n-1}), ...cl_3),cl_2),cl_1)$ is valid if the resolvents can be computed for each resolution. We, however, consider it a single order of resolution.*

Example 3.18: To understand the single sequence of resolution, let us take the following clauses.

Cl_1: W(Z, X) ←P(X, Y), Q(Y, Z).
Cl_2: P(a, b) ←S(b, a).
Cl_3: S(Y, X) ←T(X).

The resolution of clauses in the present context follows a definite order: 1-2-3 (or 3-2-1). It may be noted that

$$R(R(cl_1, cl_2), cl_3) = R(R(cl_3, cl_2), cl_1) = W(Z, a) \leftarrow Q(b, Z), T(a).$$

Consequently, a single sequence is maintained in the process of resolution of multiple program clauses.

Definition 3.16: *When in a set of resolvable clauses, resolution takes place following different orders,* **multiple sequence** *is said to be present.*

Example 3.19: The following clauses are taken to illustrate the multiple sequences in a set of resolvable clauses.

Given Cl_1: S(Z, X) ←P(X, Y), Q(Y, Z), R(Z, Y).
         Cl_2: P(a, b) ←T(c, a), U(b).
         Cl_3: Q(b, c) ←V(b), M(c).
         Cl_4: R(Z, Y) ←N(Y), O(Z).

Sequence 1: Order: 1-2-3-4.

$R(cl_1, cl_2)$:S(Z, a) ←Q(b, Z), R(Z, b), T(c, a), U(b). $|s_{12=\{a/X, b/Y\}}$
$R(R(cl_1, cl_2), cl_3)$: S(c, a) ←R(c, b), T(c, a), U(b), V(b), M(c). $|s_{12,3=\{c/Z\}}$

$R(R(R(cl_1, cl_2), cl_3), cl_4)$:
$S(c, a) \leftarrow T(c, a), U(b), V(b), M(c), N(b), O(c). \mid s_{12,3,4=\{c/Z,b/Y\}}$

Sequence 2: Order: 3-1-2-4.

$R(cl_3, cl_1)$: $S(c, X) \leftarrow P(X, b), R(c, b), V(b), M(c). \mid s_{31=\{c/Z, b/Y\}}$
$R(R(cl_3, cl_1), cl_2)$:
$S(c, a) \leftarrow T(c, a), U(b), P(a, b), R(c, b), V(b), M(c). \mid s_{31,2=\{a/X\}}$
$R(R(R(cl_3, cl_1), cl_2), cl_4)$:
$S(c, a) \leftarrow T(c, a), U(b), V(b), M(c), N(b), O(c). \mid s_{31,2,4=\{c/Z, b/Y\}}$

Sequence 3: Order: 1-2-4-3.

$R(R(cl_1, cl_2), cl_4)$:
$S(Z, a) \leftarrow Q(b, Z), T(c, a), U(b), N(b), O(Z). \mid s_{12,4=\{b/Y\}}$
$R(R(R(cl_1, cl_2), cl_4), cl_3)$:
$S(c, a) \leftarrow T(c, a), U(b), V(b), M(c), N(b), O(c). \mid s_{1,2,4,3=\{c/Z\}}$

The results of the above computation reveal that

$$R(R(R(cl_1, cl_2), cl_3), cl_4)$$
$$= R(R(R(cl_3, cl_1), cl_2), cl_4)$$
$$= R(R(R(cl_1, cl_2), cl_4), cl_3)$$
$$= S(c, a) \leftarrow T(c, a), U(b), V(b), M(c), N(b), O(c).$$

Consequently, multiple orders exist in the resolution of clauses.

Readers may please note that resolution between cl_2 and cl_3, cl_3 and cl_4, cl_2 and cl_4 are not possible.

Definition 3.17: *Let $S = \{cl_1, cl_2, ..., cl_n\}$ be a set of resolvable clauses and the cl_is are ordered in a manner that cl_i and cl_{i+1} are resolvable for $i = 1, 2, ..., (n-1)$. If cl_n is also resolvable with cl_1 we call it **circular resolution**.* Circular resolution is not allowed as it invites multiple resolutions between two clauses.

Example 3.20: Consider the following propositional clauses.

Cl_1: $Q \leftarrow P$.
Cl_2: $R \leftarrow Q$.
Cl_3: $P \leftarrow R$.

Let $Cl_{12} = R(Cl_1, Cl_2)$
$\qquad\quad = R \leftarrow P$.

However, evaluation of $R(Cl_{12}, Cl_3)$ cannot be performed as it invites more than one resolution between the two clauses[2].

Definition 3.18: *Let S be a set of resolvable clauses such that their pair-wise selection for SLD resolution from S is random. Under this condition S is called the set of an **order-less** or **order independent** clauses.*

Example 3.21: Let S be the set of the following clauses:

Cl_1: U ←P, Q, R.
Cl_2: S, M, P ←V.
Cl_3: Q ←W, M, T.
Cl_4: T ←U, S.

In this example, we attempt to resolve each clause with others.

Cl_{12}: U, S, M ←V, Q, R.
Cl_{13}: U ←P, R,W, M, T.
Cl_{14}: T ←P, Q, R, S.
Cl_{23}: S, P, Q ←V, W, T.
Cl_{34}: Q ←W, M, U, S.
Cl_{24}: M, P, T ←V, U.

As each of the clauses is resolvable with each other, order-less condition for resolution holds here.

3.3 Concurrency in Resolution

To speed up the execution of logic programs, we in this section take a look at possible parallelism/concurrency in the resolutions involved in the program.

3.3.1 Preliminary Definitions

Definition 3.19: *If S includes multiple ordered sequence of clauses for SLD resolution and for each such sequence the final resolvent is identical then the clauses in S are **concurrently resolvable**. Under this case resolution of all the clauses can be done concurrently yielding same resolvent.*

Example 3.22: In this example we consider concurrent resolution of propositional clauses.

    Cl_1: R ←P, Q.
    Cl_2 : P ←S.
    Cl_3 : Q ←T.
    Cl_4 : T ←U.

Sequence 1: Order: 1-2-3-4.

    $R(Cl_1, Cl_2)$: R ←Q, S.
    $R(R(Cl_1, Cl_2), Cl_3)$: R ←T, S.
    $R(R(R(Cl_1, Cl_2), Cl_3), Cl_4)$: R ←U, S.

Sequence 2: Order: 3-1-4-2.

    $R(Cl_3, Cl_1)$: R ←P, T.
    $R(R(Cl_3, Cl_1), Cl_4)$: R ←P, U.
    $R(R(R(Cl_3, Cl_1), Cl_4), Cl_2)$: R ←U, S.

Sequence 3: Order: 3-1-2-4.

    $R(R(Cl_3, Cl_1), Cl_2)$: R ←S, T.
    $R(R(R(Cl_3, Cl_1), Cl_2), Cl_4)$: R ←U, S.

Sequence 4: Order: 3-4-1-2.

    $R(Cl_3, Cl_4)$: Q ←U.
    $R(R(Cl_3, Cl_4), Cl_1)$: R ←P, U.
    $R(R(R(Cl_3, Cl_4), Cl_1), Cl_2)$: R ←U, S.

Illustration of concurrent resolution:

    R ← P, Q.
    P ← S.
    Q ← T.
    <u>T ← U.</u>
    R ← U, S.

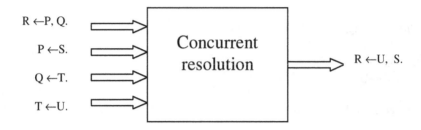

$R \leftarrow P, Q.$

$P \leftarrow S.$

$Q \leftarrow T.$

$T \leftarrow U.$

Concurrent resolution

$R \leftarrow U, S.$

Fig. 3.2: Concurrent resolution

For all the four sequences the resolvent is unique 'R ←U, S.', which is illustrated vide Fig. 3.2.

Definition 3.20: *In case multiple sequences exist in orderly resolution and the resolvent is not unique, then the substitutions used in resolution in each pair of clauses may be propagated downstream in the process of SLD resolution. The composition of the substitution sets for every two sequential substitutions is also carried forward along the SLD tree until the final resolvent is obtained. The final substitution may now be used as the instantiation space of the resolvent and the resulting clause thus generated for each such sequence is compared. In case the instantiated resolvent generated following multiple sequences yields a unique result then the clauses in S are also called* **concurrently resolvable set of resolution**. *The final substitution set is called the* **deferred substitution set**.

Example 3.23: To illustrate the aforementioned situation, let

Cl_1: R(Z, X) ←P(X, Y), Q(Y, Z).
Cl_2: P(a, b) ←S(b, a).
Cl_3: Q(b, c) ←T(c, b).
Cl_4: T(Z, Y) ←U(X, Y).

Sequence 1: Order: 1-2-3-4.

$R(Cl_1, Cl_2)$: R(Z, a) ←Q(b, Z), S(b, a). $|$ $s_{12=\{a/X, b/Y\}}$
$R(R(Cl_1, Cl_2), Cl_3)$: R(c, a) ←T(c, b), S(b, a). $|$ $s_{12,3=\{c/Z\}}$
$R(R(R(Cl_1, Cl_2), Cl_3), Cl_4)$:R(c, a) ←S(b, a), U(X, b). $|$ $s_{12,3,4=\{c/Z,b/Y\}}$

∴Composition of the substitutions, $s_{12} \Delta s_{12,3} = \{a/X, b/Y, c/Z\}$ and the final composition of the substitutions, $s_{12} \Delta s_{12,3} \Delta s_{12,3,4} = \{a/X, b/Y, c/Z\}$.

Sequence 2: Order: 3-1-2-4.

R(Cl_3, Cl_1): R(c, X) ←P(X, b), T(c, b). | $s_{31=\{b/Y, c/Z\}}$
R(R(Cl_3, Cl_1), Cl_2): R(c, a) ←T(c, b), S(b, a). | $s_{31,2=\{a/X\}}$
R(R(R(Cl_3, Cl_1), Cl_2), Cl_4):R(c, a) ←S(b, a), U(X, b). | $s_{31,2,4=\{c/Z, b/Y\}}$

∴Composition of the substitutions, $s_{31} \Delta s_{31,2}$ = {a/X, b/Y, c/Z} and the final composition of the substitutions, $s_{31} \Delta s_{31,2} \Delta s_{31,2,4}$ = {a/X, b/Y, c/Z}.

Sequence 3: Order: 3-4-1-2

R(Cl_3, Cl_4): Q(b, c) ←U(X, b). | $s_{34=\{b/Y, c/Z\}}$
R(R(Cl_3, Cl_4), Cl_1): R(c, X) ←P(X, b), U(X, b). | $s_{34,1=\{b/Y, c/Z\}}$
R(R(R(Cl_3, Cl_4), Cl_1), Cl_2):R(c, a) ←S(b, a), U(a, b). | $s_{34,1,2=\{a/X\}}$

∴Composition of the substitutions, $s_{34} \Delta s_{34,1}$ = {b/Y, c/Z} and the final composition of the substitutions, $s_{34} \Delta s_{34,1,2} \Delta s_{34,1,2}$ = {a/X, b/Y, c/Z}.

Now, if we compute the deferred substitution set for the three sequences, we find it to be equal, the value of which is given by s ={a/X, b/Y, c/Z}. When the resolvents are instantiated by this deferred substitution set, they become equal and the final resolvent is given by R(c, a) ←S(b, a), U(a, b).

3.3.2 Types of Concurrent Resolution

There are three types of concurrent resolution:

    (1) Concurrent resolution of a rule with facts,
    (2) Concurrent resolution of multiple rules,
    (3) Concurrent resolution of both multiple rules and facts.

Various well-known types of parallelisms are involved in the concurrent resolutions.

Whenever more than one variable in a rule are instantiated with the constants, *Unification-Parallelism* takes place.

Definition 3.21: *When the predicates of a rule are attempted for matching with predicates contained in the facts, **concurrent resolution of the rule with facts** is said to take place.*

It can occur in two ways. In the first case, the literals present in the body part of a clause (AND-literals) may be searched against the literals present in the heads of the available facts, which is a special case of *AND-Parallelism.*

Example 3.24: To illustrate AND-parallelism let us consider the following clauses:

Mother(Z, Y) ←Father(X, Y), Married-to(X, Z).   (3.26)
Father(r, n) ←.   (3.27)
Married-to(r, t) ←.   (3.28)

Here, predicates in the heads of the facts given by the clauses 3.27 and 3.28 are concurrently resolved with the predicates in the body of the rule given by 3.26 yielding the resolvent

Mother(t, n) ←.

Again, when a literal present in the body of one rule may be searched concurrently against the literals present in the heads of more than one fact *OR-Parallelism* is invoked.

Example 3.25: We can illustrate OR-parallelism with the help of the following clauses:

Son(X, Y) ←Father(Y, X).   (3.29)
Father(r, n) ← .   (3.30)
Father(n, a) ← .   (3.31)

Here, the variables present in the body of the rule given by equation number 3.29 can be matched concurrently with the arguments in the heads of the facts given by equation numbers 3.30 and 3.31.

Definition 3.22: *When more than one rule are resolved concurrently in a given set of resolvable clauses, we say that* **concurrent resolution of multiple rules** *has taken place.*

Example 3.26: The following clauses are considered to explain the concurrent resolution of multiple rules:

Mother(Z, Y) ←Father(X, Y), Wife(Z, X).   (3.32)
Wife(Y, X) ←Female(Y), Married-to(Y, X).   (3.33)
Married-to(X, Y) ←Marries(X, Y).   (3.34)

Here, the concurrent resolution can take place between the rules 3.32 and 3.33 in parallel with the rules 3.33 and 3.34. Moreover, the above three rules can be concurrently resolved together yielding the resolvent

Mother(Z, Y) ←Father(X, Y), Female(Z), Marries(X, Z). (3.35)

A special kind of parallelism, known as *Stream-parallelism* can be encountered while discussing concurrent resolution of multiple rules. It is explained with the example 3.27.

Example 3.27: Let us consider the following clauses:

Integer(X+1) ←Integer(X). (3.36)
Evaluate-square(Z.Z) ←Integer(Z). (3.37)
Print(Y) ←Evaluate-square(Y). (3.38)
Integer(0) ←. (3.39)

Here, the resolvent obtained by resolving the rule 3.36 and the fact 3.39 is propagated to resolve the rule 3.37. The resolvent thus obtained is resolved further with 3.38. The process is repeated in a streamline for the integer sequence 0, 1, 2, ... up to infinity. After one result is printed, the pipeline becomes busy rest of the time, and resolution of three pairs of clauses take place.

Definition 3.23: *If more than one rule is resolved concurrently with more than one fact, concurrent resolution of both multiple rules and facts takes place.*

Example 3.28: To illustrate the concurrent resolution of multiple rules and facts, let us take the following clauses:

R(Z, X) ←P(X, Y), Q(Y, Z). (3.40)
P(a, b) ←. (3.41)
Q(b, c) ←. (3.42)
S(U, V), T(U, V) ←R(U, V). (3.43)
←S(d, e). (3.44)
←T(d, e). (3.45)

Here, concurrent resolution can take place in several ways. At first, the rules given by (3.40) and (3.43) can resolve generating the resolvent S(Z, X),T(Z, X) ←P(X, Y), Q(Y, Z). Again, the rule (3.40) is resolved with the facts given by (3.41) and (3.42) in parallel while the rule (3.43) is resolved with the facts given by (3.44) and (3.45). The resolvents in these two cases are, respectively:

R(c, a) ←Q(b, c). and ←R(d, e).

When the predicates present in the body of one rule does also occur in the head of a second rule the latter and the former rules are said to be in pipeline [9]. Rules in pipeline are resolvable. But in case there exists matching ground clause for the common literals of both the rules, it is preferred to resolve the ground clause with either of (or both) the rules. The Petri net model for extended logic programming that we would like to introduce shortly is designed based on the aforementioned concept.

The observations which can be made from example 3.28 are as follows:

(1) Resolution of a rule with one or more facts provides a scope for yielding intermediate ground inferences.

For instance, the rule (3.40) in example 3.28 when resolved with (3.41) and (3.42) yields a ground intermediate R (c, a) ←Q (b, c).

(2) Resolution of two or more rules yields new rules containing literals with renamed variables. The effort in doing so, on many occasions, may be fruitless.

For instance, if rule (3.40) and (3.43) were resolved, a new rule would be generated with no further benefits of re-resolving the resulting rule with available facts.

(3) In case there exist concurrently resolvable group of clauses, where each group contains a rule and a few facts, then the overall computational speed of the system can be significantly improved.

Example 3.29 provides an insight to this issue.

Example 3.29: Let us take the following clauses

$$R(Z, X) \leftarrow P(X, Y), Q(Y, Z). \tag{3.46}$$
$$P(a, b) \leftarrow. \tag{3.47}$$
$$Q(b, c) \leftarrow. \tag{3.48}$$
$$S(U, V), T(U, V) \leftarrow R(U, V). \tag{3.49}$$
$$\leftarrow S(c, a). \tag{3.50}$$
$$\leftarrow T(c, a). \tag{3.51}$$

Unlike example (3.28), where the resulting resolvents after concurrent resolution of two groups of clauses could not participate in further resolution, here the resolvents due to concurrent resolution of (3.46 − 3.48) and (3.49 − 3.51) can also take part in the resolution game, resulting in a null clause. To have an idea of speed-up, we construct a graph (vide Fig. 3.3) indicating the concurrency in resolution.

It is apparent from the graph (Fig. 3.3) that the concurrent resolution of clauses (3.46-3.48) and (3.49-3.51) can take place in one unit time, and the resolution of the resulting clauses require one unit time. Thus, the time taken for execution of the logic program on a parallel engine is two unit times. The same problem, if solved by SLD tree, takes as many as five unit times to perform five resolutions of binary clauses (Fig. 3.4).

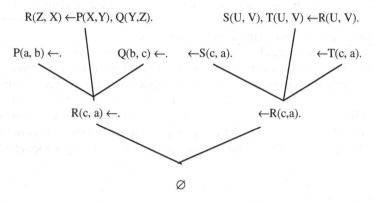

Fig. 3.3: A graph illustrating concurrency in resolution

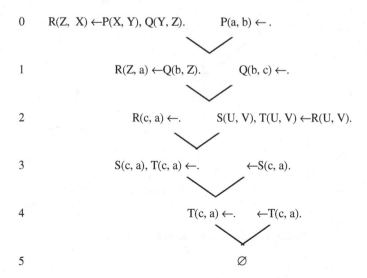

Fig. 3.4: The SLD Tree

The example shows that definitely there is a scope in speed-up due to concurrent resolution at the cost of additional expenses for hardware resources.

The most important problem in concurrent resolution is the identification of the clauses that participate in the resolution process. When there exist groups of concurrently resolvable clauses, search cost to detect the participating clauses in each group sometimes is too high to be amenable in real time. A specialized data structure, capable of performing concurrent resolution of multiple groups of clauses, thus is recommended. In fact, we are in search of a suitable structure where participating facts and rules under one group of resolvable clauses can be represented by neighboring structural units. The search cost needed in concurrent resolution thus can be saved by the above mentioned data structure.

Petri nets which have already proved itself successful in solving many complex problems of knowledge engineering, can equally be used in the present context to efficiently handle the problems of concurrent resolution. Let us for example consider a clause 'P (X, Y), Q (X, W) ←R (X, Y), S (Y, W).', which is represented in a Petri net by a transition and four associated places, where P and Q are represented by output places, and R and S are denoted by input places of the transition. The argument of each literal in a rule is represented by a specialized

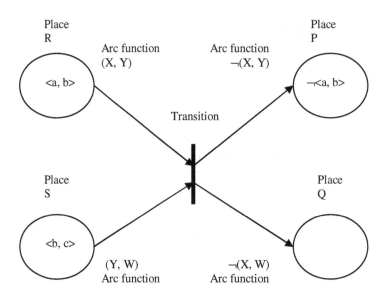

Fig. 3.5: Mapping of a rule on Petri net

function, called arc function, which is associated with the arc connecting the transition with the respective places. The arc functions are needed for generation of variable bindings in the process of resolution of clauses. If '¬P (a, b) ←.', 'R (a, b) ←.' and 'S (b, c) ←.' are supplied as additional facts then they could be mapped in the places connected with the proposed transition, and the arguments ¬<a, b>, <a, b> and<b, c> of the facts are saved as tokens of the respective places P, R and S. Such neighbourhood mapping of the rule and facts described in Fig. 3.5 help concurrent resolution with no additional time for searching the concurrently resolvable clauses.

Reasoning in Logic program with Petri nets was first proposed by Murata [11]. In this chapter, we extended Murata's model on the following counts.

♦ *In Murata's model arc functions associated with the arcs of a Petri net are positive irrespective of the type[1] of the arcs. The model to be proposed shortly, however, assigns a positive sign to the arc function attached with a place-to-transition connective arc, and a negative sign to the arc function attached with a transition-to-place connective arc. The attachment of sign with the arc functions facilitates the scope of matching of signed tokens of the respective places with the arc functions of the connected arcs following the formalisms of predicate logic.*

♦ *Unlike Murata's model, where the arguments of body-less clauses were also represented as arc functions, in the present model these are represented as tokens of the appropriate places. Thus in the present model, we can save additional transition firings for all those arc functions corresponding to the body-less clauses.*

♦ *The extension of Murata's model presented here allows AND-, OR-, Unification- and Stream-parallelism in a logic program.*

The chapter takes into account the aforementioned features and presents a new algorithm for automated reasoning, capable of handling parallelisms in a logic program. The definitions which are needed to design the algorithm for automated reasoning are given in the following section.

3.4. Petri net Model for Concurrent Resolution

This section provides a new algorithm for concurrent resolution of program clauses using a specialized model of extended Petri net.

---

[1] The directed arcs in a Petri net denote connectivities from: (i) places to transitions, and (ii) transitions to places, and thus are of two basic types.

3.4.1 Extended Petri Net

Definition 3.24: *An **Extended Petri Net (EPN)**, which will be used here for reasoning with a First Order Logic (FOL) program, is a 9-tuple, given by*

$$EPN = \{P, Tr, D, f, m, A, a, I, O\}$$

where

$P = \{p_1, p_2, ..., p_m\}$ *is a set of places;*
$Tr = \{tr_1, tr_2, ..., tr_n\}$ *is a set of transitions;*
$D = \{d_1, d_2, ..., d_m\}$ *is a set predicates;*
$P \cap Tr \cap D = \varnothing$; *Cardinality of P = Cardinality of D;*
$f: D \rightarrow P^{\infty}$ *represents a mapping from the set of predicates to the set of places;*
$m: P \rightarrow <x_i, ...,y_i, X, ...,Y,f, ...,g >$ *is an association function, represented by the mapping from places to terms, which may include signed constant(s) like x_i, ..., y_i, variable(s) like X, ...,Y and function f, ...,g of variables;*
$A \subseteq (P \times Tr) \cup (Tr \times P)$ *is the set of arcs, representing the mapping from the places to the transitions and vice versa;*
$a: A \rightarrow (X, Y, ..., Z)$ *is an association function of the arcs, represented by the mapping from the arcs to terms. For arcs $A \in (P \times Tr)$ the arc functions a are positively signed, while for arcs $A \in (Tr \times P)$ the arc functions a are negatively signed;*
$I: Tr \rightarrow P^{\infty}$ *is a set of input places, represented by the mapping from the transitions to their input places;*
$O: Tr \rightarrow P^{\infty}$ *is a set of output places, represented by the mapping from the transitions to their output places.*

3.4.2 Mapping a Clause onto Extended Petri Net

Consider a first order clause cl_i where the arguments of the predicates contain variables/constants only.

$$Cl_i: p_{m+1}(Z, X), \;.....,p_{m+n}(Z, Y) \leftarrow p_1(X, Y), \;....., p_m(Y, Z).$$

To construct a Petri net corresponding to the above clause we use the following procedure.

Procedure Petri net construction for a rule

Input: A given first order clause with predicates containing variables and constants.

Output: An extended Petri net.

Begin

1. Construct a transition and label it as tr_i corresponding to clause cl_i.

2. Check whether any place p_k, $\exists k$ already exist in the so far constructed net. If yes, construct places p_1 through p_m excluding p_k. Else draw m places and arcs emanating from those places to the transitions. Attach labels p_1, p_2, ..., p_m to designate the places. Attach the argument of the predicate p_i as the arc function in the arc connected between place p_i and the given transition. Repeat it for all i from 1 to m.

3. Check whether any place p_j already exists in the so far constructed net. If yes, construct places p_{m+1} through p_{m+1} through p_{m+n} excluding p_j. Else construct n number of places and connect these places from the transition by outgoing arcs. Label the places as p_{m+1}, p_{m+2}, ..., p_{m+n}. Attach negated argument of predicate p_j in the arc connected between the given transition and place p_j. Repeat at for all j between m+1 to m+n.

End.

3.4.3 Mapping a Fact onto Extended Petri Net

Procedure mapping fact onto EPN

Begin
    If the predicate name used to denote the fact is already existing in a given EPN
    Then
        If the fact occurs in the body of the given clause
        Then the negated argument of the clause is used as token and inserted into the corresponding place of the EPN.
        Else the positive argument of the clause is used as token and inserted into the corresponding place of the EPN.
        Endif
    Endif
End.

The algorithms introduced above can be used to map all the rules and facts on to an EPN.

Example 3.30: Mapping of the aforementioned parameters onto an EPN is illustrated (vide Fig. 3.6) in this example with the following FOL clauses:

$$\text{Son}(Y, Z), \text{Daughter}(Y, Z) \leftarrow \text{Father}(X, Y), \text{Wife}(Z, X). \tag{3.52}$$
$$\text{Father}(r, l) \leftarrow . \tag{3.53}$$
$$\text{Wife}(s, r) \leftarrow . \tag{3.54}$$
$$\neg \text{Daughter}(l, s) \leftarrow . \tag{3.55}$$

Here, $P = \{p_1, p_2, p_3, p_4\}$;
   $Tr = \{tr_1\}$;
   $D = \{d_1, d_2, d_3, d_4\}$ with $d_1 =$ Father, $d_2 =$ Wife, $d_3 =$ Son and $d_4 =$ Daughter;
   $f(\text{Father}) = p_1$, $f(\text{Wife}) = p_2$, $f(\text{Son}) = p_3$, $f(\text{Daughter}) = p_4$;
   $m(p_1) = <r, l>$, $m(p_2) = <s, r>$, $m(p_3) = < \varnothing >$, $m(p_4) = \neg<l,s>$ initially and
   can be computed subsequently by resolution of clauses on the EPN;
   $A = \{A_1, A_2, A_3, A_4\}$, and
   $a(A_1) = (X, Y)$, $a(A_2) = (Z, X)$, $a(A_3) = \neg(Y, Z)$, $a(A_4) = \neg(Y, Z)$ are the arc
   functions;
   $I(tr_1) = \{p_1, p_2\}$, and $O(tr_1) = \{ p_3, p_4 \}$.

It is to be noted that if-then operator of the knowledge has been represented in the Fig. 3.6 by tr_1 and the antecedent-consequent pairs of knowledge have been denoted by input(I)-output(O) places of tr_1. Moreover, the arguments of the predicates have been represented by arc functions.

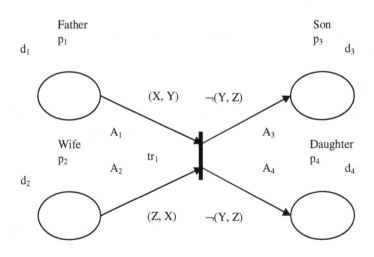

Fig. 3.6: Parameters of an EPN used to represent knowledge in FOL

3.5 Concurrent Resolution on Petri Nets

The enabling and firing conditions of transitions are explained below to describe the resolution principles on an EPN.

3.5.1 Enabling and Firing Conditions of a Transition

In order to determine the enabling and firing conditions of a transition we need to define the following items:

Consistent bindings
Let $a_j \; \exists$ be an arc function of arc A_j associated with a transition tr_i in an EPN, and $m(p_j)$ denotes the token at place p_j where

$$A_j \in \{p_j\} \times \{tr_i\} \cup \{tr_i\} \times \{p_j\}.$$

1. *If number of elements in a_j is equal to that of $m(p_j)$, then we assign the kth element of $m(p_j)$ to $a_j \; \forall k$.*

2. *The assignment of step 1 is repeated for all j.*

3. *Let X, Y, ..., Z be the list of variables present in $\cup a_j$, $\forall j$. If X = k is present in all excluding at most one a_j, then variable X is said to have a consistent value.*

   *If all the variables X, Y, ..., Z have consistent values, then we say that the arc function variables associated with a transition have **consistent bindings**.*

Current-bindings (c-b) *denote the set of instantiation of all the variables associated with the transitions.*

Used-bindings (u-b) *denote the set of union of the current-bindings up to the last transition firing.*

Properly signed token *means tokens with proper signs, i.e., positive tokens for input places and negative tokens for output places of a transition.*

Inactive arc functions *represent the arc functions associated with a transition, which do not participate in the process of generation of consistent bindings of variables.*

Inert place *represent the place connected with the inactive arc function.*

Enabling Conditions: *A transition is **enabled**, if i) all excluding at most one inert place associated with the transition tr_i possesses properly signed tokens and ii) the*

*variables associated with the arc functions of the transition have consistent
bindings.*

Firing Condition: *A transition is **fired** if it is enabled and the current-bindings is
not a subset of used-bindings.*

3.5.2 Algorithm for Concurrent Resolution

The algorithm for automated reasoning to be presented shortly allows concurrent
firing of multiple transitions. The algorithm in each pass checks the enabling
conditions of all the transitions. If one or more transitions are found enabled, the
unifier (here referred to as current-bindings) for each transition is searched against
the union of the preceding unifiers (used-bindings) of the said transition. If the
current-bindings are not members of the respective used-bindings, then the
enabled transitions are fired concurrently. The tokens for the inert places are then
computed following the current-bindings of the fired transitions.

One question that naturally arises: how long should we continue firing of the
transitions? The firing may continue until no new inferences are derived. This is
taken into account in the procedure Automated-Reasoning.

Procedure Automated-reasoning

Begin
    For each transition do
        Par Begin
        used-bindings:= Null;
        Flag:= true; // transition not firable.//
            Repeat
                If (a transition is enabled) AND (current-bindings is not a member of
                used-bindings)
                Then do Begin
                Fire the transition and send tokens to the ***inert place***
                using the set current-bindings and following the
                inactive arc function with a presumed opposite sign;
                 Update used-bindings by taking union with current-bindings;
                Flag:= false; //record of transition firing//
                Increment no-of-firing by 1;
                End
                Else Flag:= true;
            End;
            Until no transition fires;
        Par End;
End.

The algorithm is now verified taking the following rules with the help of Fig. 3.7.

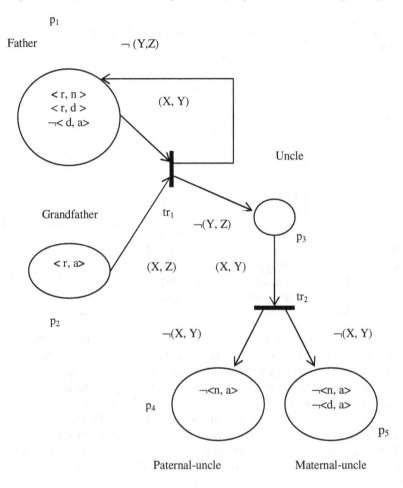

Fig. 3.7: An illustrative Petri net with initial assignment of tokens used to verify the procedure automated-reasoning

Rules:

Father (Y, Z), Uncle (Y, Z) ←Father (X, Y), Grandfather (X, Z).	(3.56)
Paternal-uncle (X, Y), Maternal-uncle (X, Y) ←Uncle (X, Y).	(3.57)
Father (r, n) ←.	(3.58)
Father (r, d) ←.	(3.59)
¬Father (d, a) ←.	(3.60)
Grandfather (r, a) ←.	(3.61)
¬ Paternal-uncle (n, a) ←.	(3.62)
¬Maternal-uncle (n, a) ←.	(3.63)
¬Maternal-uncle(d, a) ←.	(3.64)

Table 3.1: Trace of the algorithm on example net of Fig. 3.7.

Time slot	Tran.	Set of c-b	Set of u-b	Flag = 0, if c-b $\notin$ u-b = 1, if c-b $\in$ u-b $\neq \{\phi\}$ or c-b=$\{\phi\}$
First cycle	tr_1	$\{r/x,d/y,a/z\}$	$\{\{\phi\}\}$	0
	tr_2	$\{n/x,a/y\}$	$\{\{\phi\}\}$	0
Second cycle	tr_1	$\{r/x,n/y,a/z\}$	$\{\{r/x,d/y,a/z\}\}$	0
	tr_2	$\{d/x,a/y\}$	$\{\{n/x, a/y\}\}$	0
Third cycle	tr_1	$\{r/x,d/y,a/z\}$ / $\{r/x,n/y,a/z\}$	$\{\{r/x,d/y,a/z\}, \{r/x,n/y,a/z\}\}$	1
	tr_2	$\{n/x, a/y\}$/ $\{d/x, a/y\}$	$\{\{n/x, a/y\}, \{d/x, a/y\}\}$	1

The Petri net shown in Fig. 3.7 is constructed with a set of rules (3.56 – 3.64). The reasoning algorithm presented earlier is then invoked and the trace of the algorithm thus obtained is presented in Table 3.1. It is clear from the table that the current-bindings (c-b) are not members of used bindings (u-b) in the first two Firing Criteria Testing (FCT) iterations. Therefore, flag = 0. Thus following the algorithm, transitions tr_1 and tr_2 both fire concurrently. In the third iteration current-bindings become members of used-bindings, and consequently flag = 1; so no firing takes place during the third iteration. Further, number of transition in the Petri net (vide Fig. 3.7) being two only, control exits the repeat-until loop in procedure Automated-reasoning after two FCT iterations.

3.5.3 Properties of the Algorithm

Theorem 1 is provided to demonstrate that the proposed algorithm includes all types of parallelisms and theorem 2 shows the completeness of the same.

THEOREM 1: *The procedure Automated-reasoning supports AND, OR and Stream-parallelisms.*

Proof: Let

$$p_i \in I(tr_k) \text{ for } i = 1,2,.....,m \tag{3.65}$$

and $p_o \in O(tr_k)$ for o =m+1, m+2,....., m+n. (3.66)

$p_j \in I(tr_{k+1})$ for $\exists j$, where $\{p_j\} \cap \{p_o\} \neq \varnothing$, i.e., there exists some common place between the output place of transition tr_k and input place of transition tr_{k+1}. So the two transitions tr_k and tr_{k+1} are in pipeline.

Let $p_l \in \{p_j\} \cap \{p_o\}$ for $\exists l$.

From (A) and (B), we obtain the general clause represented by the predicates mapped at the input and the output places of the transition tr_k. If d_i denotes the predicates corresponding to place p_i $\forall I$, then the rule under consideration is given by (C).

$$d_{m+1}(.), \ldots, d_{m+n}(.) \leftarrow d_1(.), \ldots, d_m(.). \tag{3.67}$$

Now assume that the knowledge base includes the following excluding at most one fact.

$d_1(.) \leftarrow .$
$d_2(.) \leftarrow .$
.
.
$d_m(.) \leftarrow .$
$\neg d_{m+1}(.) \leftarrow .$
$\neg d_{m+2}(.) \leftarrow .$
.
.
$\neg d_{m+n}(.) \leftarrow .$

Further, let tr_{k+l} be another transition where for $\exists u$, $p_u \in I(tr_k) \cap O(tr_k) \cap I(tr_{k+1})$, i.e., p_u is a common input-output place of the transition tr_k. Without any loss of generality, let us assume that, $p_1 = p_{m+n} = p_u$. In case new stream of tokens are generated by p_1-tr_k-p_{m+n}, then the transition tr_{k+1} under favorable condition of consistent binding will also fire concurrently with transition tr_k but on new sets of data. Thus stream-parallelism if exists in the logic program, can definitely be realized with our proposed algorithm.

Note: It is to be noted that unification-parallelism can always be maintained in a Petri net model with additional resources for concurrent instantiation of variables with tokens.

Example 3.31: If we consider the clauses given in (3.56-3.64), we can find out that AND-parallelism takes place when (3.59) and (3.61) are resolved together with (3.56). OR-parallelism takes place when (3.58) and (3.59) are tried for resolution together with (3.56). Also OR-parallelism takes place when (3.63) and

(3.64) are resolved together with (3.57). In each case whenever a variable is being matched with a constant or another variable, unification–parallelism takes place.

In this book we consider Petri net models capable of representing multiple antecedent and multiple consequent clauses. Usually the commas present in the antecedent clauses denote conjunction and in consequent clauses denote disjunction. Thus in presence of tokens at all but one input/output places of an enabled transition, the transition will fire generating a new token. Such firing of transition includes typical AND- and a different type of OR-parallelism. Here, independent facts mapped at the output places of the transition behave like typical OR-clauses, and a set of concurrent resolution takes place between the OR-clauses and a given rule containing those literals present in the OR-clauses as consequents.

Unification-parallelism can always be maintained in the Petri net model, and Stream-parallelism exists only when the network includes pipelined transitions where a transition in the pipeline waits for the other to generate a sequence of tokens.

THEOREM 2: *The procedure Automated-reasoning is complete.*

Proof: When a transition in procedure Automated-reasoning fires, concurrent resolution takes place among a number of tokens with the main clause represented by the transition and its input-output places. It is important to note that the concurrent resolution mentioned above is similar with a number of binary resolutions of clauses following SLD resolution technique. Secondly when two transitions having common input places are enabled they can fire, since the concurrent firing of transitions in the present context is conflict-free. Thus firing of one transition does not present another to fire. Thirdly, after firing of one transition, tokens are entered into an inert place connected with the transition, and the old tokens in the places associated with the transition are not removed. This ensures that the tokens non utilised in one transition firing may be utilized in subsequent firing of the same and other transition(s). Consequently, all possible inferences that can be derived from the given set of clauses using SLD resolution, can also be derived by procedure Automated-reasoning. Since SLD resolution is complete, the proof of completeness of the procedure Automated–reasoning naturally follows.

3.6 Performance Analysis of Petri Net-based Models

In this section we outline two important issues: the ***speed-up factor*** and the ***resource utilization rate*** of the proposed algorithm when realized on a parallel architecture.

3.6.1 The Speed-up

A complexity analysis of a logic program of n clauses comprising of predicates of arity p reveals that the time T_u required for execution of the program by SLD-resolution on a uniprocessor architecture is given by

$$T_u = O(p.n). \tag{3.68}$$

The product (p.n) in the order of complexity appears because of SLD-resolution of n clauses with p sequential matching of arguments of predicates involved in the resolution process.

The same program comprising of m_1, m_2, ..., m_k number of concurrently resolvable clauses is executed[2] on a pipelined (multiprocessor) architecture, capable of resolving max $\{m_i: 1 \le i \le k\}$ number of clauses in a unit time. Let m_i include s_i number of supplied clauses and d_i number of derived clauses. Thus $\Sigma m_i = \Sigma s_i + \Sigma d_i$. Under this circumstance the total computational time T_m for execution of the logic program is given by

$$T_m = O[p(n - (s_1 + s_2 + s_3 + \dots + s_k) - 1 + 1 \times k)]$$

$$= O[p(n - \Sigma_{i=1}^{k} s_i + k-1)]$$

$$\approx O[p(n - \Sigma_{i=1}^{k} s_i + k)] \tag{3.69}$$

When Σs_i approaches Σm_i, T_m is maximum.

The above result presumes a k-stage pipeline of the k-sets of concurrently resolvable clauses. If the k-sets of clauses are independent, then the concurrent resolution of all the k-set of clauses can be accomplished within a unit time, and thus the computational complexity further reduces to $O[p(n - \Sigma_{i=1}^{k} s_i)]$. Thus irrespective of a program, it can easily be ascertained that the computational time T_m of a typical logic program always lies in the interval:

$$O[p(n - \Sigma_{i=1}^{k} s_i)] \le T_m \le O[p(n - \Sigma_{i=1}^{k} s_i + k)] \tag{3.70}$$

---

[2] The k sets of concurrent clauses here is assumed to be resolved in sequence, i.e., resolution of one concurrent set of clause is dependent on a second set, and thus all sets of concurrent clauses cannot be resolved in parallel.

Thus, *Speed-up*[3] in the worst case is found to be

$$S = T_u/T_m$$

$$= (p.n)/ [p(n - \Sigma_{i=1}^{k} s_i +k)]$$

$$= n / [n- \Sigma_{i=1}^{k} s_i + k)] \tag{3.71}$$

In case all the n number of program clauses are exhausted by resolution, i.e. $\Sigma_{i=1}^{k} m_i$ approaches n, then S is maximized, and the speed-up factor, S_{max} is given by

$$S_{max} = n / k. \tag{3.72}$$

The last expression reveals that smaller is the k, larger is the S_{max}. The best case corresponds to k =1, when there is a single set of concurrently resolvable clauses. But since $\Sigma_{i=1}^{k} s_i = n$ and k =1 in the present context, $\Sigma_{i=1}^{k} s_i = \Sigma_{i=1}^{1} s_i = s_1 = n$, which means all the n set of clauses are resolvable together. Consequently the speed-up factor is n.

On the other extreme end, when k = n, i.e., there are s_1, s_2, ..., s_n number of concurrently resolvable sets of clauses, then $S_{max} = n/n = 1$, and there is no speed-up. In fact this case corresponds to typical SLD-resolution and the number of clauses $s_1 = s_2 = ... = s_n = 2$.

3.6.2 The Resource Utilization Rate

Let us assume that the number of resources available for concurrent resolution in the present context is max $\{s_k : 1 \leq k \leq n\}$. Thus maximum degree of parallelism [5] P is given by

$$P = max \{s_k : 1 \leq k \leq n\}. \tag{3.73}$$

The average degree of parallelism P_{av} is defined below following Hwang and Briggs [5] as

$$P_{av} = (\Sigma_{i=1}^{k} s_i)/ k. \tag{3.74}$$

---

[3] In case of unification-parallelism (realized in our architecture, vide chapter 4), T_m reduces to $O[(n- \Sigma_{i=1}^{k} s_i + k)]$ and consequently the worst-case speed-up factor becomes $S_{max} = (p. n) / (n- \Sigma_{i=1}^{k} s_i + k)$.

The *Resource Utilization Rate* μ thus is found to be

$$\mu = P_{av} / P$$

$$= (\Sigma_{i=1}^{k} s_i)/ \; [k. \max \{s_k : 1 \leq k \leq n\}]. \tag{3.75}$$

When s_i for all i approaches to max $\{s_k : 1 \leq k \leq n\}$, $\Sigma_{i=1}^{k} s_i = k. \max\{s_k: 1 \leq k \leq n\}$, and consequently, μ approaches 1.

3.6.3 Resource Unlimited Speed-up and Utilization Rate

Suppose the number of resources $\geq$ n, the no. of program clauses. Then the concurrent resolution of different sets of clauses may take place in parallel. Suppose, out of s_1, s_2, ...,s_k number of concurrent sets of resolvable clause, r-sets of clauses on an average can participate in concurrent resolutions at the same time. Then the average time T_{RU} required to execute the program $= O[p \; (n- \Sigma_{i=1}^{k} s_i + k/r)]$. Then, *Resource Unlimited Speed-up*

$$S_{RU} = T_u / T_{RU}$$

$$= (p.n)/ \; [p(n - \Sigma_{i=1}^{k} s_i + k/r)]$$

$$= n / (n- \Sigma_{i=1}^{k} s_i + k/r) \tag{3.76}$$

Consequently, maximum speed-up occurs when $\Sigma_{i=1}^{k} s_i$ approaches n, and the result is

$$(S_{RU})_{max} = (n/k)r. \tag{3.77}$$

Further, maximum degree of parallelism in a resource unlimited system is $P_{RU} = \Sigma_{i=1}^{k} s_i$, and the average degree of parallelism $P_{av} = \Sigma_{i=1}^{r} s_i$. Thus *Resource Utilization Rate* is given by

$$\mu = P_{av} / P_{RU}$$

$$= (\Sigma_{i=1}^{r} s_i)/ (\Sigma_{i=1}^{k} s_i) \tag{3.78}$$

In a special case, when $s_i = s$ for all i = 1 to k, the above ratio reduces to (r / k). It is to be noted that when more than one group of concurrently resolvable clauses participate in the resolution process at the same time, $\max\{s_k: 1 \leq k \leq n\}$ assumes the maximum of the sum of the concurrently resolvable group of clauses.

Example 3.32: Let s_1 and s_5 be two groups of clauses that can independently be resolved at the same time. Further, if we assume that $s_1 + s_5 \geq s_2, s_3, s_4, \ldots$ Consequently $\max\{s_k\}$ will take the value $s_1 + s_5$.

3.7 Conclusions

The chapter presented a new algorithm for automated reasoning in a logic program using extended Petri net models. Because of the structural advantage of Petri net models, the proposed algorithm is capable of handling AND-, OR-, unification- and stream-parallelisms in a logic program. A complexity analysis of a logic program with n number of clauses and k sets of concurrently resolvable clauses reveals that the maximum speed-up factor of the proposed algorithm in the worst case is O (n/k). Under no constraints on resources, the speed-up factor is improved further by an additional factor of r, where r denotes an average number of the concurrent sets of resolvable clauses. In absence of any constraints to resources, the maximum resource utilization rate for the proposed algorithm having $s_1 = s_2 = \ldots = s_n$ is O(r/k). With limited resource architecture, the proposed algorithm can execute safely at the cost of extra computational time. The selection of dimensions of such limited resource architecture depends greatly on the choice of r. Selection of r in typical logic programs, in turn, may be accomplished by running Monte Carlo simulations for a large set of programs. A complete study of this, however, is beyond the scope of the present work.

With the increasing use of logic programs in data modeling, its utility in the next generation commercial database systems will also increase in pace. Such systems require specialized engine that supports massive parallelisms. The proposed computational model, being capable of handling all possible parallelisms in a logic program, is an ideal choice for exploration in commercial database systems. To meet up this demand, a hardware realization of the proposed algorithm is needed. The next chapter of the thesis extends the theoretical EPN model of automated reasoning to a specialized MIMD architecture of a central processing unit to be explored in the next generation database machines.

Exercises

1. Given below a set of clauses:

    (i) $r \vee s \leftarrow p \wedge q$.
    (ii) R(Z, f(X)) ←P(X, Y), Q(Y, Z).
    (iii) Above(a, c) ←Above(a, b), Above(b, c).
    (iv) N(f(Y), X) ←L(X, Y, Z), M(f(X), Z).

(a) Identify the literals in the head part,
(b) Identify the literals in the body part,
(c) List the functions,
(d) List the arguments of the head literals,
(e) List the arguments of the body literals.

[**Answer:**

(a) The literals in the head part:
(i)   r, s;
(ii)   R(Z, f(X));
(iii)   Above(a, c);
(iv)   N(f(Y), X).

(b) The literals in the body part:
(i)   p, q;
(ii)   P(X, Y), Q(Y, Z);
(iii)   Above(a, b), Above(b, c);
(iv)   L(X, Y, Z), M(f(X), Z).

(c)   The functions:
(i)   Nil;
(ii)   f(X);
(iii)   Nil;
(iv)   f(Y), f(X).

(d)   The arguments of the head literals:
(i)   Nil;
(ii)   Z, f(X);
(iii)   a, c;
(iv)   f(Y), X.

(e)   The arguments in the body literals:
(i)   Nil;
(ii)   X, Y and Y, Z;
(iii)   a, b and b, c;
(iv)   (X, Y, Z) and f(X), Z.]

2. Identify the following from the given set of clauses

(i)   ground literal,
(ii)   goal clause or query,
(iii)   fact,

(iv)   Horn clause,
(v)   Non-Horn clause,
(vi)   Propositional clause.

(a)   Boy(X) ←Male-child(X), Non-adult(X).
(b)   Boy(X), Girl(X) ←Male-child(X), Non-adult(X).
(c)   S ←P, Q, R.
(d)   Boy(X) ←.
(e)   Boy(ram) ←.
(f)   ←Boy(Y).

[**Answer:**

(i)   Ground literal: Boy(ram)

(ii) Goal clause or query: ←Boy(Y).

(iii) Fact: Boy(ram) ←.

(iv) Horn clause: Boy(X) ←Male-child(X), Non-adult(X).
                          S ←P, Q, R.
                  Boy(X) ←.
                  Boy(ram) ←.
                          ←Boy(Y).

(v) Non-Horn clause: Boy(X), Girl(X) ←Male-child(X), Non-adult(X).

(vi) Propositional clause: S ←P, Q, R.]

3. Translate the following clauses into English:

        Mother (Z, Y) ←Father (X, Y) ∧ Wife (Z, X).
        Mother (s, l) ←.
        Wife (s, r) ←.
        Query: ←Father (r, l).

[**Hints:** If X is the father of Y and Z is the wife of X, then Z is the mother of
        Y.
        s is the mother of l.
        s is the wife of r.
        Whether r is the father of l?]

4. Construct an Extended Logic Program from the following statements:

The animals that eat plants are herbivorous, the animals that eat animals are carnivorous and the animals that eat plants and animals both are omnivorous.

[**Hints:** Herbivorous (X) ←Animals (X)∧Plant (Y)∧Eats (X, Y).
Carnivorous (X) ←Animals (X)∧Plant (Z)∧Eats (X, Z).
Omnivorous (X) ←Animals (X)∧Plant (Y)∧Animals (Z)∧
Eats (X, Y)∧Eats (X, Z).]

5. Given an expression
$$w = P (X, f (Y, Z), d)$$
and three substitution sets
$$s_1 = \{a/X, b/Y, c/Z\}$$
$$s_2 = \{g(Y)/X\}$$
$$s_3 = \{g(W)/X, b/Y, c/Z\}.$$
Evaluate ws_1, ws_2 and ws_3.

[**Answer:** ws_1 = P(a, f(b, c), d),
ws_2 = P(g(Y), f(Y, Z), d) and
ws_3 = P(g(W), f(b, c), d).]

6. Let w be an expression and ws be an expression after substitution, what is the substitution set?

$$w: M(Z, Y) ←F(X, Y), K(Z, X).$$
$$ws: M(s, j) ←F(r, j), K(s, r).$$

[**Answer:** s:$\{r/X, j/Y, s/Z\}$.]

7. Let s_1 and s_2 be two substitutions such that
$$s_1: \{ r/X \}, \text{ and}$$
$$s_2: \{u(X)/Y\}.$$

Evaluate the composition of the substitutions:
(a) $s_1 \Delta s_2$,
and (b) $s_2 \Delta s_1$.

[**Answer:** The composition of the substitutions:

(a) $s_1 \Delta s_2 = \{r/X, u(X)/Y\}$,
(b) $s_2 \Delta s_1 = \{u(r)/Y, r/X\}$.]

8. Verify the substitution set property 1, i.e., $(ws_1)\Delta s_2 = w(s_1 \Delta s_2)$ using the following items:

> Let the expression w: P(X, Y, Z)
> And the substitution sets s_1: {u(X)/Y, v(X)/Z},
> $\quad\quad\quad\quad\quad\quad\quad$ s_2: {r/X}.

[**Answer:** $(ws_1)\Delta s_2 = $ P(X, u(X), v(X)){r/X}
$\quad\quad\quad\quad\quad\quad$ = P(r, u(r), v(r)).

$\quad\quad$ $w(s_1\Delta s_2) = $ (P(X, Y, Z)){r/X, u(r)/Y, v(r)/Z}
$\quad\quad\quad\quad\quad\quad\quad$ = P(r, u(r), v(r)).
$\quad\quad$ $\therefore$ $(ws_1)\Delta s_2 = w(s_1\Delta s_2).$]

9. Verify the substitution set property 2, i.e., $(s_1\Delta s_2)\Delta s_3 = s_1\Delta(s_2\Delta s_3)$ using the following items:

> Let the substitution sets s_1: {r/X },
> $\quad\quad\quad\quad\quad\quad\quad\quad$ s_2: {u(X)/Y} and
> $\quad\quad\quad\quad\quad\quad\quad\quad$ s_3: {l/Z}.

[**Answer:** $s_1\Delta s_2 = $ {r/X, u(X)/Y}
$\quad\quad$ $(s_1\Delta s_2)\Delta s_3 = $ {r/X, u(X)/Y, l/Z}.

$\quad\quad\quad\quad$ $s_2\Delta s_3 = $ {u(X)/Y, l/Z}
$\quad\quad$ $s_1\Delta(s_2\Delta s_3) = $ {r/X, u(X)/Y, l/Z}.
$\quad\quad$ $\therefore (s_1\Delta s_2)\Delta s_3 = s_1\Delta(s_2\Delta s_3).$]

10. Verify the substitution set property 3, i.e., $s_1\Delta s_2 \neq s_2\Delta s_1$ using the following items:

> Let the substitution sets s_1: {r/X }, and
> $\quad\quad\quad\quad\quad\quad\quad\quad$ s_2: {u(X)/Y}.

[**Answer:**
$\quad\quad\quad\quad$ $s_1\Delta s_2 = $ {r/X, u(X)/Y} and
$\quad\quad\quad\quad$ $s_2\Delta s_1 = $ {u(r)/Y, r/X}.
$\quad\quad$ $\therefore$ $s_1\Delta s_2 \neq s_2\Delta s_1.$]

11. Which of the following clauses are resolvable and what is the resolvent?

(i) Cl_1: $\quad$ R(Y, X) ←P(X, Y).
(ii) Cl_2: $\quad$ P(r, n) ← .
(iii) Cl_3: $\quad$ R(n, r) ←.

[**Hints:** As the same predicate is present in the head and the body part of the clause number cl_2 and cl_1 respectively, they are resolvable whereas due to presence of the same predicate in the head part of the clauses, cl_1 and cl_3, they are not resolvable.]

12. By definition 3.9, show that the following is a set of resolvable clauses.

    (i) Cl_1:   R(Z, X) ←P(X, Y), Q(Y, Z).
    (ii) Cl_2:   P(r, s) ← .
    (iii) Cl_3:   ¬R(l, r) ←.
    (iv) Cl_4:   Q(s, l) ←.

    [**Hints:** Cl_{12}:   R(Z, r) ←Q(s, Z).
                    Cl_{13}:   ←P(r, Y), Q(Y, l).
                    Cl_{14}:   R(l, X) ←P(X, s).

    As each of the clauses is resolvable with at least one of the set of clauses, producing a resolvent, according to definition 3.9, it is a set of resolvable clauses.]

13. Identify the definite program clause with definite goal.

    Cl_1: R, S ←P, Q.
    Cl_2: R ←P, Q.
    Cl_3: ←P, Q.

    [**Hints:** According to definition 3.10, Cl_2: R ←P, Q. is a definite clause containing one atom in its head and Cl_3: ←P, Q. is a definite goal with empty consequent (see definition 3.12).]

14. Construct the resolution tree for linear selection,

    R ←P, Q.
    S ←R.
    T ←S.
    Q ←.
    P ←.

    Goal: ←T.

    [**Answer:** The resolution tree is constructed vide Fig. 3.8.

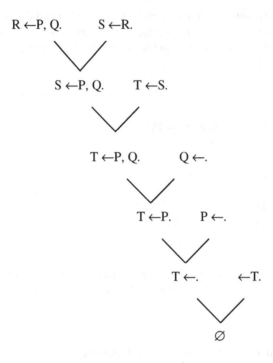

Fig. 3.8: The SLD resolution tree]

15. Determine whether the following are orderly or order independent clauses? If orderly, verify whether single or multiple sequence? Determine the various orders of resolution in the following set of clauses.

Cl_1: R(Z, X) ←P(X, Y), Q(Y, Z).
Cl_2: P(r, a) ←.
Cl_3: Q(a, k) ←.
Cl_4: ¬R(k, r) ←.

[**Hints:** The clauses are to be selected pair-wise according to some definite order from the set of resolvable clauses. Otherwise they fail to generate a solution. As for example, cl_2, cl_3 or cl_3, cl_4 cannot be resolved. But, we can get results by resolving the clauses following some definite orders.

The multiple orders are:

Sequence 1: Order 1: 1-2-3-4/ 2-1-3-4
$Cl_{12\text{-}3\text{-}4}$: Ø

Sequence 2: Order 2: 1-2-4-3/ 2-1-4-3
$Cl_{12\text{-}4\text{-}3}$: ∅

Sequence 3: Order 3: 1-3-2-4/ 3-1-2-4
$Cl_{13\text{-}2\text{-}4}$: ∅

Sequence 4: Order 4: 1-3-4-2/ 3-1-4-2
$Cl_{13\text{-}4\text{-}2}$: ∅

Sequence 5: Order 5: 1-4-2-3/ 4-1-2-3
$Cl_{14\text{-}2\text{-}3}$: ∅

Sequence 6: Order 6: 1-4-3-2/ 4-1-3-2
$Cl_{14\text{-}3\text{-}2}$: ∅]

16. Determine the final resolvent from the given set of clauses. If not possible, indicate why.

   Cl_1: Son(Y, X) ←Father(X, Y).
   Cl_2: Mother(i, a) ←Son(a, n), Husband(n, i).
   Cl_3: Father(Z, Y) ←Mother(X, Y), Wife(X, Z).

[Hints: Here,

   Cl_{12}: Mother(i, a) ←Father(n, a), Husband(n, i).
   $Cl_{12\text{-}3}$ is not possible as double resolution takes place.]

17. Show that the following clauses are order independent. Explain the reason for non-validity.

   Cl_1: Boy(X), Girl(X) ←Child(X).
   Cl_2: Likes-to-play-indoor(X) ←Boy(X), Introvert(X).
   Cl_3: Likes-to-play-doll(X) ←Girl(X), Likes-to-play-indoor(X),
                                        Has-doll(X).

   [Hints: Each clause is resolvable with any other clause.
            Non-validity: Ultimately double resolution takes place.]

18. Show whether the following set of clauses are concurrently resolvable.

   Cl_1: R(Z, X) ←P(X, Y), Q(Y, Z).
   Cl_2: P(r, a) ←.
   Cl_3: Q(a, k) ←.
   Cl_4: ¬R(k, r) ←.

[**Hints:** As the instantiated resolvent generated following multiple sequences yields a unique result, the set includes a concurrent set of resolution.]

19. Test whether the following clauses are concurrently resolvable.

(a) Cl_1: Mother(Z, Y) ←Father(X, Y), Son(Y, Z), Married-to(X, Z).
     Cl_2: Father(r, l) ←Has-one-son(r, l), Male(r).
     Cl_3: Female(Z) ←Mother(Z, Y).
     Cl_4: Son(l, s) ←.

(b) Cl_1: Mother(Z, Y) ←Father(X, Y), Son(Y, Z), Married-to(X, Z).
     Cl_2: Father(r, l) ←Has-one-son(r, l), Male(r).
     Cl_3: Female(Z) ←Mother(Z, Y).
     Cl_4: Son(k, s) ←.

[**Hints:**

(a) As the set includes multiple ordered sequence of clauses for SLD resolution and for each such sequence the final resolvent is identical, then the orderly resolution is concurrent resolution.

(i) **Sequence 1:** Order 1: 1-2-3-4.

> Cl_{12}: Mother(Z, l) ←Son(l, Z), Has-one-son(r, l), Male(r),
>   Married-to(r, Z).
> $Cl_{12\text{-}3}$: Female(Z) ←Son(l, Z), Has-one-son(r, l), Male(r),
>   Married-to(r, Z).
> $Cl_{12\text{-}3\text{-}4}$: Female(s) ←Has-one-son(r, l), Male(r), Married-to(r, s).

(ii) **Sequence 2:** Order 2: 1-2-4-3.

> Cl_{12}: Mother(Z, l) ←Son(l, Z), Has-one-son(r, l), Male(r),
>   Married-to(r, Z).
> $Cl_{12\text{-}4}$: Mother(s, l) ←Has-one-son(r, l), Male(r),
>   Married-to(r, s).
> $Cl_{12\text{-}4\text{-}3}$: Female(s) ←Has-one-son(r, l), Male(r), Married-to(r, s).

(iv) **Sequence 3:** Order 3: 1-3-4-2.

> Cl_{13}: Female(Z) ←Father(X, Y), Son(Y, Z), Married-to(X, Z).
> $Cl_{13\text{-}4}$: Female(s) ←Father(X, l), Married-to(X, s).
> $Cl_{13\text{-}4\text{-}2}$: Female(s) ←Has-one-son(r, l), Male(r), Married-to(r, s).

(v) **Sequence 4:** Order 4: 1-3-2-4.

        Cl_{13}: Female(Z) ←Father(X, Y), Son(Y, Z), Married-to(X, Z).
        Cl_{13-2}: Female(Z) ←Male(r), Has-one-son(r, l), Son(l, Z),
                Married-to(r, Z).
        Cl_{13-2-4}: Female(s) ←Male(r), Has-one-son(r, l), Son(l, s),
                Married-to(r, s).

(vi) **Sequence 5:** Order 5: 1-4-2-3.

        Cl_{14}: Mother(s, l) ←Father(X, l), Married-to(X, s).
        Cl_{14-2}: Mother(s, l) ←Has-one-son(r, l), Male(r), Married-to(r, s).
        Cl_{14-2-3}: Female(s) ←Has-one-son(r, l), Male(r), Married-to(r, s).

(vii) **Sequence 6:** Order 6: 1-4-3-2.

        Cl_{14}: Mother(s, l) ←Father(X, l), Married-to(X, s).
        Cl_{14-3}: Female(s) ←Father(X, l), Married-to(X, s).
        Cl_{14-3-2}: Female(s) ←Has-one-son(r, l), Male(r), Married-to(r, s).

(b) As the final resolvent is not identical, the clauses are not concurrently
     resolvable.

(i) **Sequence 1:** Order 1: 1-2-3-4.

        Cl_{12}: Mother(Z, l) ←Son(l, Z), Has-one-son(r, l), Male(r),
                Married-to(r, Z).
        Cl_{12-3}: Female(Z) ←Son(l, Z), Has-one-son(r, l), Male(r),
                Married-to(r, Z).
        Cl_{12-3-4}: Not possible.

(ii) **Sequence 2:** Order 2: 1-2-4-3.

        Cl_{12}: Mother(Z, l) ←Son(l, Z), Has-one-son(r, l), Male(r),
                Married-to(r, Z).
        Cl_{12-4}: Not possible.

(iii) **Sequence 3:** Order 3: 1-3-4-2.

        Cl_{13}: Female(Z) ←Father(X, Y), Son(Y, Z), Married-to(X, Z).
        Cl_{13-4}: Female(s) ←Father(X, k), Married-to(X, s).
        Cl_{13-4-2}: Not possible.

(iv) **Sequence 4:** Order 4: 1-3-2-4.

Cl_{13}: Female(Z) ←Father(X, Y), Son(Y, Z), Married-to(X, Z).
Cl_{13-2}: Female(Z) ←Male(r), Has-one-son(r, l), Son(l, Z),
        Married-to(r, Z).
Cl_{13-2-4}: Not possible.

(v) **Sequence 5:** Order 5: 1-4-2-3.

Cl_{14}: Mother(s, k) ←Father(X, k), Married-to(X, s).
Cl_{14-2}: Not possible.

(vi) **Sequence 6:** Order 6: 1-4-3-2.

Cl_{14}: Mother(s, k) ←Father(X, k), Married-to(X, s).
Cl_{14-3}: Female(s) ←Father(X, k), Married-to(X, s).
Cl_{14-3-2}: Not possible.

So, we can easily find out that the clauses are not concurrently resolvable in
case of (b) as there is no unique result after the resolution following different
sequences.]

20. Indicate (a) the AND-parallel clauses and (b) the OR-parallel clauses, to
concurrently resolve the rule cl_1 with the rest of the clauses cl_2 through cl_5.

Cl_1: Likes-to-play(X, Y) ←Child(X), Game(Y).
Cl_2: Child(r) ←.
Cl_3: Child(t) ←.
Cl_4: Game(c) ←.
Cl_5: Game(l) ←.

[**Answer:**

(a) The AND-parallel clauses:
    (i) Cl_1, Cl_2, Cl_4,
    (ii) Cl_1, Cl_2, Cl_5,
    (iii) Cl_1, Cl_3, Cl_4,
    (iv) Cl_1, Cl_3, Cl_5.

(b) The OR-parallel clauses:
    (i) Cl_1, Cl_2 and Cl_1, Cl_3,
    (ii) Cl_1, Cl_4 and Cl_1, Cl_5.]

21. (a) Verify whether concurrent resolution is valid for the following clauses:

Cl_1: Game(Y) ←Child(X), Likes-to-play(X, Y).
Cl_2: Outdoor-game(Y), Indoor-game(Y) ←Game(Y).
Cl_3: Child(X) ←Boy(X).
Cl_4: Child(X) ←Girl(X).
Cl_5: Boy(X) ←Likes-to-play(X, Y), Outdoor-game(Y).
Cl_6: Girl(X) ←Likes-to-play(X, Y), Indoor-game(Y).
Cl_7: Likes-to-play(t, c) ←.
Cl_8: Likes-to-play(r, l) ←.
Cl_9: Outdoor-game(c) ←.
Cl_{10}:¬ Indoor-game(c) ←.

(b) If yes, then identify the types of the concurrent resolution for the following sequences:

(i) Cl_5, Cl_7, Cl_9,
(ii) Cl_1, Cl_2, Cl_5,
(iii) Cl_3, Cl_5, Cl_7, Cl_9,
(iv) Cl_1, Cl_2, Cl_6,
(v) Cl_1, Cl_2, Cl_3, Cl_7.

[**Hints:**

(a) For the following ordered sequence of clauses for SLD resolution, the final resolvent is identical. So, concurrent resolution is possible in the following cases:

(i)   **Sequence 1:** Order 1: 1-3-7.
       Sequence 2: Order 2: 1-7-3.

(ii)   **Sequence 3:** Order 3: 1-3-8.
       Sequence 4: Order 4: 1-8-3.

(iii) **Sequence 5:** Order 5: 1-4-7.
       Sequence 6: Order 6: 1-7-4.

(iv)   **Sequence 7:** Order 7: 1-4-8.
       Sequence 8: Order 8: 1-8-4.

(v)   **Sequence 9:** Order 9: 5-8-9.
       Sequence 10: Order 10: 5-9-8.

(b)
  (i)   Concurrent resolution between a rule and facts,
  (ii) Concurrent resolution between rules,
  (iii) Concurrent resolution between rules and facts,
  (iv) Concurrent resolution between rules,
  (v)   Concurrent resolution between fact and rules.]

22. Map the following FOL clauses on to an EPN and state the result after
resolution.

Cl_1: Game(Y) ←Child(X), Likes-to-play(X, Y).
Cl_2: Outdoor-game(Y), Indoor-game(Y) ←Game(Y).
Cl_3: Child(t) ←.
Cl_4: Likes-to-play(t, c) ←.
Cl_5:¬ Indoor-game(c) ←.

[**Hints:** An EPN is constructed with the FOL clauses vide Fig. 3.9.

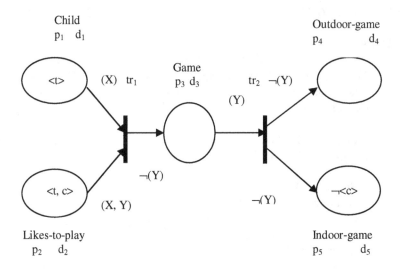

Fig. 3.9: Mapping on to an Extended Petri net

Here, P = {p_1, p_2, p_3, p_4, p_5};
Tr = {tr_1, tr_2};
D = {d_1, d_2, d_3, d_4, d_5};
f(Child) = p_1, f(Likes-to-play) = p_2, f(Game) = p_3,
f(Outdoor-game) = p_4, f(Indoor-game) = p_5;
m(p_1) = < t >, m(p_2) = <t, c>, m(p_3) = <Ø>, m(p_4) = <Ø>,
m(p_5) = ¬< c > initially and can be computed in the process of resolution;
A = {A_1, A_2, A_3, A_4, A_5, A_6}; and
a(A_1) = (X), a(A_2) = (X, Y), a(A_3) = ¬(Y), a(A_4) = (Y), a(A_5) = ¬(Y) and a(A_6) = ¬(Y) are the arc functions;
I(tr_1) = {p_1, p_2}, I(tr_2) = {p_3};
O(tr_1) = {p_3}, O(tr_2) = {p_4, p_5}.

Resolution between the rules cl_1, cl_2 and the facts cl_3, cl_4, cl_5 takes place concurrently yielding resulting token at the place 'Outdoor-game': '<c>'.]

23. Map the given program clauses onto an EPN. What result do you obtain   after execution of the program by the algorithm: 'Procedure Automated-Reasoning'.

The program clauses are:

Cl_1: Reproduce-by-laying-eggs(X) ←Build-nests(X), Lay-eggs(X).
Cl_2: Has-wings(X) ←Can-fly(X), Has-feather(X).
Cl_3: Bird(X) ←Reproduce-by-laying-eggs(X), Has-beaks(X),
                   Has-wings(X).
Cl_4: Build-nests(p) ←.
Cl_5: Lay-eggs(p) ←.
Cl_6: Can-fly(p) ←.
Cl_7: Has-feather(p) ←.
Cl_8: Has-beaks(p) ←.

[**Hints:** An EPN vide Fig. 3.10 is constructed with the given program clauses.

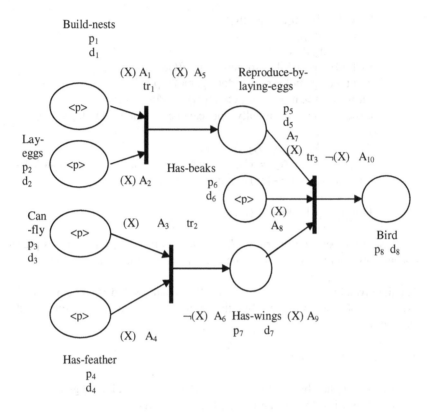

Fig. 3.10: The Extended Petri Net illustrating the problem

Here, $P = \{p_1, p_2, p_3, p_4, p_5, p_6, p_7, p_8\}$;
$Tr = \{tr_1, tr_2, tr_3\}$;
$D = \{d_1, d_2, d_3, d_4, d_5, d_6, d_7, d_8\}$;
$f(\text{Build-nests}) = p_1$, $f(\text{Lay-eggs}) = p_2$, $f(\text{Can-fly}) = p_3$,
$f(\text{Has-feather}) = p_4$, $f(\text{Reproduce-by-laying-eggs}) = p_5$,
$f(\text{Has-beaks}) = p_6$, $f(\text{Has-wings}) = p_7$, $f(\text{Bird}) = p_8$;
$m(p_1) = \text{<p>}$, $m(p_2) = \text{<p>}$, $m(p_3) = \text{<p>}$, $m(p_4) = \text{<p>}$, $m(p_5) = \text{<\varnothing>}$,
$m(p_6) = \text{<p>}$, $m(p_7) = \text{<\varnothing>}$, $m(p_8) = \text{<\varnothing>}$ initially;
$A = \{A_1, A_2, A_3, A_4, A_5, A_6, A_7, A_8\}$;
$a(A_1) = (X)$, $a(A_2) = (X)$, $a(A_3) = (X)$, $a(A_4) = (X)$, $a(A_5) = \neg(X)$,
$a(A_6) = \neg(X)$, $a(A_7) = (X)$, $a(A_8) = (X)$, $a(A_9) = (X)$, $a(A_{10}) = \neg(X)$;
$I(tr_1) = \{p_1, p_2\}$, $I(tr_2) = \{p_3, p_4\}$, $I(tr_3) = \{p_5, p_6, p_9\}$;
$O(tr_1) = \{p_5\}$, $O(tr_2) = \{p_7\}$, $O(tr_3) = \{p_8\}$.

At first the transitions tr_1 and tr_2 will fire satisfying the three conditions required for transition firing. Then after firing of tr_3, the resultant token will be obtained at the place p_8 : <p>.]

24. Given a set of clauses(Cl_1-Cl_{13}):

      Cl_1: Father(Y, Z), Uncle(Y, Z) ←Father(X, Y), Grandfather(X, Z).
      Cl_2: Paternal-uncle(X, Y), Maternal-uncle(X,Y) ←Uncle(X, Y).
      Cl_3: Mother(Z, Y) ←Father(X, Y), Married-to(X, Z).
      Cl_4: Father(X, Y) ←Mother(Z, Y), Married-to(X, Z).
      Cl_5: Father <r, n> ←.
      Cl_6: Father <r, d> ←.
      Cl_7: ¬ Father <d, a> ←.
      Cl_8: Grandfather <r, a> ←.
      Cl_9: ¬Paternal-uncle <n, a> ←.
      Cl_{10}: ¬Maternal-uncle <n, a> ←.
      Cl_{11}: ¬Maternal-uncle <d, a> ←.
      Cl_{12}: Married-to <r, t> ←.
      Cl_{13}: Married-to <n, i> ←.

(a) List the possible resolutions that take place in the network.
(b) Identify the concurrent resolutions among those in the above list.
(c) Represent the concurrent resolutions in tabular form like Table 2.1.

[Hints:
(a) The possible resolutions are:

      1-3, 1-4, 1-5, 1-6, 1-7, 1-8, 2-9, 2-10, 2-11, 3-12, 3-13, 4-12, 4-13.

(b) The concurrent resolutions are:

Sequence 1: Order 1: 1-6-7-8.
    $Cl_{1-6-7-8}$ = Uncle(d, a) ←.

Sequence 2: Order 2: 2-9-10.
    Cl_{2-9-10} = ←Uncle(n, a). ≡ ¬ Uncle(n, a) ←.

Sequence 3: Order 3: 3-5-12.
    Cl_{3-5-12} = Mother(t, n) ←.

Sequence 4: Order 4: 3-6-12.
    Cl_{3-6-12} = Mother(t, d) ←.

(c) The concurrent resolutions are given in Table 3.2.

Table 3.2: Trace of the algorithm on example net of Fig. 3.12.

Time slot	Trans.	Set of c-b	Set of u-b	Flag = 0, if c-b $\notin$ u-b =1, if c-b$\in$ u-b $\neq \{\varnothing\}$ or c-b = $\{\varnothing\}$
First cycle	tr_1	{r/X, d/Y, a/Z}	{{$\varnothing$}}	0
	tr_2	{n/X, a/Y}	{{$\varnothing$}}	0
	tr_3	{r/X, n/Y, t/Z}/ {r/X, d/Y, t/Z}	{{$\varnothing$}}	0
Second cycle	tr_1	{r/X, n/Y, a/Z}	{{r/X, d/Y, a/Z}}	0
	tr_2	{d/X, a/Y}	{{n/X, a/Y}}	0
	tr_3	{$\varnothing$}	{{r/X, n/Y, t/Z}, {r/X, d/Y, t/Z}}	0
Third cycle	tr_1	{$\varnothing$}	{{r/X, d/Y, a/Z}}	1
	tr_2	{$\varnothing$}	{{n/X, a/Y}}	1
	tr_3	{n/X, a/Y, i/Z}	{{r/X, n/Y, t/Z}, {r/X, d/Y, t/Z}, {n/X, a/Y, i/Z}}	1

]

25. (a) For answering the goal(←M(t, n).), draw the SLD-tree from the given set of clauses,

   (b) Use 'Procedure Automated-Reasoning' to verify the goal,

   (c) Assuming unit time to perform a resolution, determine the computational time involved for execution of the program by SLD-tree approach,

   (d) Compute the computational time required for execution of the program clauses on EPN using 'Procedure Automated-Reasoning',

   (e) Determine the percentage time saved in the EPN approach ((a-b)/a × 100) where 'a' stands for SLD, 'b' for EPN.

[**Hints:** (a) We obtain the result as M(t, n) ←. From the SLD-tree given below vide Fig.3.11.

Mother(Z, Y) ←Father(X, Y), Married-to(X, Z).   Married-to <r, t> ←.

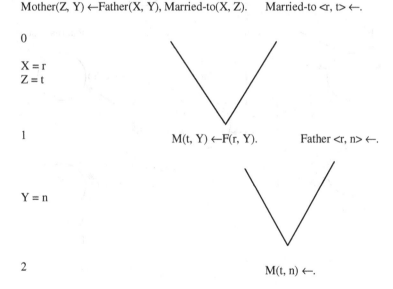

0

X = r
Z = t

1   M(t, Y) ←F(r, Y).   Father <r, n> ←.

Y = n

2   M(t, n) ←.

Fig. 3.11: The SLD-tree for answering the goal: ←M(t, n)

(b) After mapping onto the Extended Petri net as in the Fig. 3.12, we can
    easily find out that in the pass 1, all three transitions (tr_1, tr_2 and tr_3)
    satisfy the three conditions for transition firing:

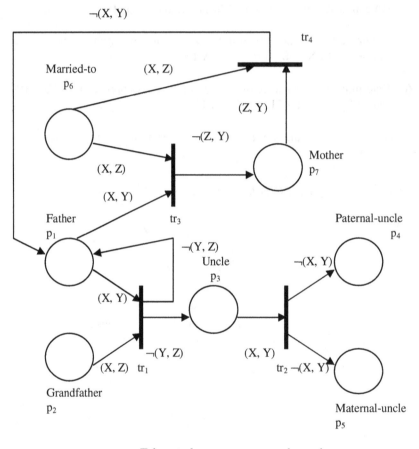

Token at place p_1 = <r, n>, <r, d>, ¬<d, a>,
Token at place p_2 = <r, a>,
Token at place p_4 = ¬<n, a>,
Token at place p_5 = ¬<n, a>, ¬<d, a>,
Token at place p_6 = <r, t>, <n, i>.

Fig. 3.12: The set of given clauses mapped onto an Extended Petri Net

(i)   All but one input and output places contain properly signed tokens,

(ii)   The variables in the arguments of the arc functions contain consistent bindings,

(iii)   The current bindings are not the sub set of the used bindings.

Therefore, the transitions tr_1, tr_2 and tr_3 fires concurrently yielding tokens at the places:

$<d, a>$ at p_3,

$\neg <n, a>$ also at p_3 and

$<t, n>, <t, d>$ at p_7.

From the above we can see that the required goal ($\leftarrow M(t, n).$) can be obtained by the algorithm 'Procedure Automated-Reasoning'.

(c) Assuming unit time to perform a resolution, the computational time involved for execution of the program by SLD-tree approach can easily be found out from the Fig. 3.11 and is found to be 2.

(d) The computational time required for execution of the program clauses on EPN (Fig. 3.12) using 'Procedure Automated-Reasoning' to obtain the required goal (as only one pass is required for one set of concurrent resolution) is found out to be 1.

(e) The percentage time saved in the EPN approach:
$((a-b)/a \times 100)$, where 'a' stands for SLD, 'b' for EPN
$= ((2-1)/2) \times 100 \%$
$= 50 \%.$]

26. (a) Show that after complete execution of the algorithm 'Procedure Automated-Reasoning' on the given EPN (Fig. 3.12) the following conclusions are obtained,

(i)   Mother(i, a) $\leftarrow$.
(ii)   Paternal-uncle(d, a) $\leftarrow$.
(iii)   Mother(t, n) $\leftarrow$.
(iv)   Mother(t, d) $\leftarrow$.

(b) Assuming the above four predicates as four independent goals, construct SLD-tree for each case,

(c) Compute the computational time involved in part (a) and part (b) and hence determine the computational time saved by EPN approach.

[Hints:

(a) After mapping on to the Extended Petri net as in the Fig. 3.12, we can easily find out that in the pass 1, all three transitions satisfy the three conditions for transition firing:

    (i)   All but one input and output places contain properly signed tokens,

    (ii)   The variables in the arguments of the arc functions contain consistent bindings,

    (iii)   The current bindings are not the subset of the used bindings.

Therefore, the transitions tr_1, tr_2 and tr_3 fire concurrently yielding tokens at the places:

For tr_1: <d, a> at p_3,
For tr_2: ¬<n, a> also at p_3 and
For tr_3: <t, n> (i.e., Mother(t, n) ←.), <t, d> (i.e., Mother(t, d) ←.) at p_7.

In the pass 2, only tr_1 and tr_2 fires satisfying the required conditions generating the tokens:

For tr_1: <n, a> at p_1,
For tr_2: <d, a> (i.e., Paternal-uncle(d, a) ←.) at p_4.
In the final pass 3, tr_3 fires only yielding the token:
For tr_3: <i, a> (i.e., Mother(i, a) ←.) at p_7.

(b) (i) The SLD-tree for generating the goal 'Mother(i, a) ←.' is constructed in Fig. 3.13.

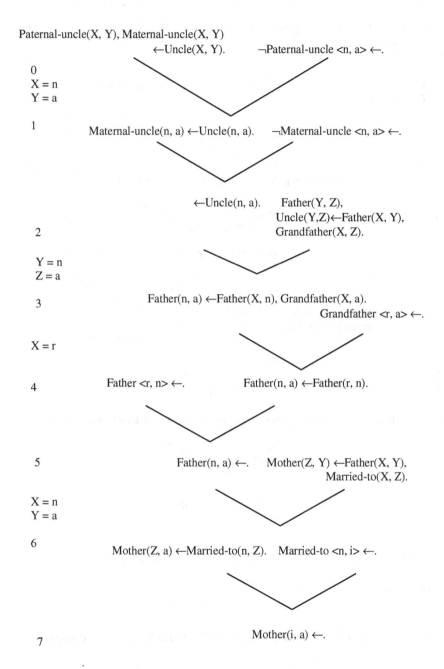

Paternal-uncle(X, Y), Maternal-uncle(X, Y)
                    ←Uncle(X, Y).   ¬Paternal-uncle <n, a> ←.

0
X = n
Y = a

1   Maternal-uncle(n, a) ←Uncle(n, a).   ¬Maternal-uncle <n, a> ←.

                          ←Uncle(n, a).   Father(Y, Z),
                                           Uncle(Y,Z)←Father(X, Y),
2   Grandfather(X, Z).

Y = n
Z = a

3   Father(n, a) ←Father(X, n), Grandfather(X, a).
                                           Grandfather <r, a> ←.

X = r

4   Father <r, n> ←.   Father(n, a) ←Father(r, n).

5   Father(n, a) ←.   Mother(Z, Y) ←Father(X, Y),
                                         Married-to(X, Z).

X = n
Y = a

6   Mother(Z, a) ←Married-to(n, Z).   Married-to <n, i> ←.

7   Mother(i, a) ←.

Fig. 3.13: The SLD-tree for generating the goal Mother(i, a) ←.

(ii) The SLD-tree for generating the goal 'Paternal-uncle(d, a) ←.' is constructed in Fig. 3.14.

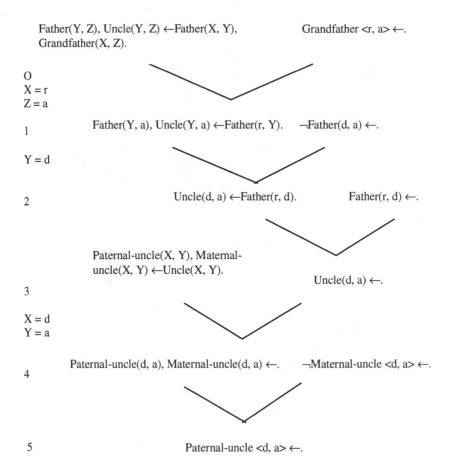

Fig. 3.14: The SLD-tree for generating the goal: Paternal-uncle(d, a) ←.

(iii) Same as the Fig.3.11.

(iv) The SLD-tree for generating the goal 'Mother(t, d) ←.' is constructed in Fig. 3.15.

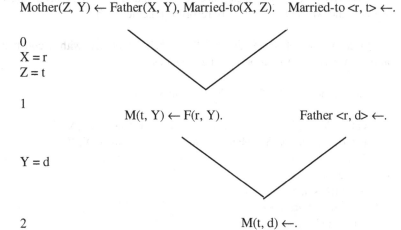

Mother(Z, Y) ← Father(X, Y), Married-to(X, Z).   Married-to <r, t> ←.

0
X = r
Z = t

1
            M(t, Y) ← F(r, Y).   Father <r, d> ←.

Y = d

2   M(t, d) ←.

Fig. 3.15: The SLD-tree for evaluating the goal M(t, d) ←.

(c) For part (a), assuming unit time to perform a resolution, computational time involved for execution of the program by SLD-tree approach will be seen as seven steps are needed according to the SLD tree shown in the Fig. 3.13.

But, for part (b), only three passes are needed for execution of the program on EPN using Procedure 'Automated Reasoning'. Therefore, the computational time required for execution of the program on EPN using 'Procedure Automated-Reasoning' is found to be three.

∴ The percentage time saved in the EPN approach:
((7-3)/7) × 100 = 57.14%.]

27. Point out in the Fig. 3.12 when the AND/OR/Unification-parallelism takes place.

[**Hints:** AND-parallelism takes place when cl_6 and cl_8 are resolved together with cl_1.

OR-parallelism takes place when cl_5 and cl_6 are tried for resolution together with cl_1.

Also OR-parallelism takes place when cl_{10} and cl_{11} are resolved together with cl_2 and when cl_{12} and cl_{13} are resolved together with cl_{14}.

In each case whenever an variable is being matched with a constant or another variable, unification-parallelism takes place.]

28. From the given clauses of Fig. 3.7 draw the SLD-tree.

Cl_1: Father(Y, Z), Uncle(Y, Z) ←Father(X, Y), Grandfather(X, Z).
Cl_2: Paternal-uncle(X, Y), Maternal-uncle(X,Y) ←Uncle(X, Y).
Cl_3: Father <r, n> ←.
Cl_4: Father <r, d> ←.
Cl_5: ¬ Father <d, a> ←.
Cl_6: Grandfather <r, a> ←.
Cl_7: ¬Paternal-uncle <n, a> ←.
Cl_8: ¬Maternal-uncle <n, a> ←.
Cl_9: ¬Maternal-uncle <d, a> ←.

[**Hints:** Here, two parallel SLD-trees will be formed vide Figs. 3.16 and 3.17.

(i)

Father(Y, Z), Uncle(Y, Z) ←Father(X, Y), Grandfather(X, Z).

Father <r, d> ←.

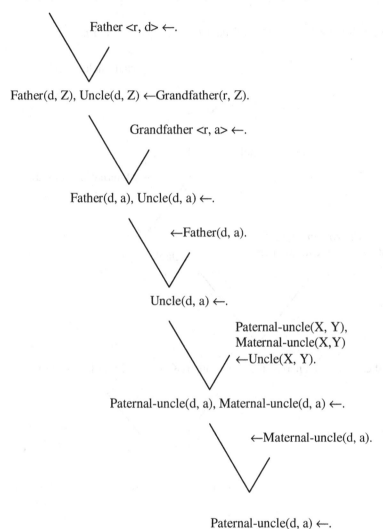

Father(d, Z), Uncle(d, Z) ←Grandfather(r, Z).

Grandfather <r, a> ←.

Father(d, a), Uncle(d, a) ←.

←Father(d, a).

Uncle(d, a) ←.

Paternal-uncle(X, Y),
Maternal-uncle(X,Y)
←Uncle(X, Y).

Paternal-uncle(d, a), Maternal-uncle(d, a) ←.

←Maternal-uncle(d, a).

Paternal-uncle(d, a) ←.

Fig. 3.16: One SLD-tree

(ii)

Paternal-uncle(X, Y), Maternal-uncle(X, Y) ←Uncle(X, Y).

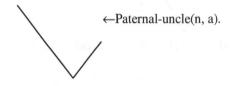

←Paternal-uncle(n, a).

Maternal-uncle(n, a) ←Uncle(n, a).

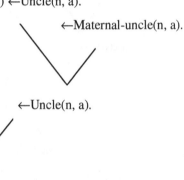

←Maternal-uncle(n, a).

Father(Y, Z), Uncle(Y, Z) ←
Father(X, Y), Grandfather(X, Z).   ←Uncle(n, a).

Father(n, a) ←Father(X, n), Grandfather(X, a).   Grandfather <r, a> ←.

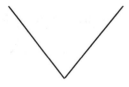

Father(n, a) ←Father(r, n).   Father(r, n) ←.

Father(n, a) ←.

Fig. 3.17: Another SLD-tree   ]

29. Represent the concurrently resolvable clauses in the EPN in Fig. 3.7 for the following clauses by a tree structure.

[**Hints:** Two parallel reasoning will take place giving rise to two parallel tree-like structures vide Fig. 3.18 and Fig. 3.19.

One tree:

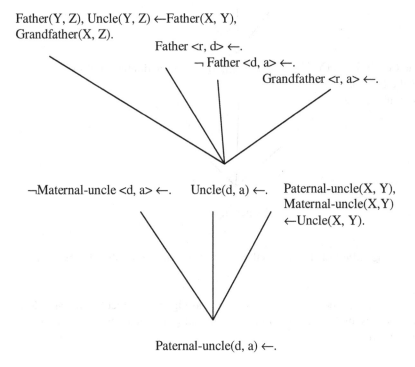

Fig. 3.18: SLD-tree for one set of concurrently resolvable clauses

Another tree:

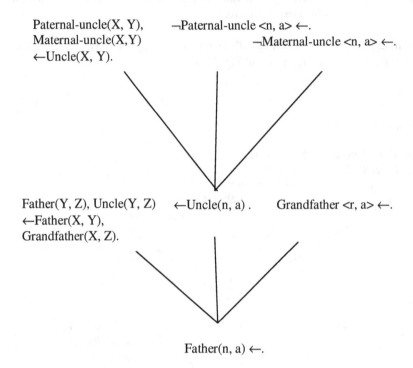

Paternal-uncle(X, Y),   ¬Paternal-uncle <n, a> ←.
Maternal-uncle(X,Y)   ¬Maternal-uncle <n, a> ←.
←Uncle(X, Y).

Father(Y, Z), Uncle(Y, Z)   ←Uncle(n, a) .   Grandfather <r, a> ←.
←Father(X, Y),
Grandfather(X, Z).

Father(n, a) ←.

Fig. 3.19: SLD-tree for another set of concurrently resolvable clauses]

30. From the given Petri nets (vide Fig. 3.20 and Fig. 3.21), calculate the speed-up
    and the Resource utilization rate from the definition given in the equation no.
    (3.71) and (3.75).

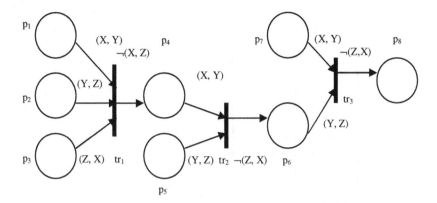

Token at the place p_1 = <a, b>, Token at the place p_2 = <b, c>,
Token at the place p_3 = <c, a>, Token at the place p_5 = <c, d>,
Token at the place p_7 = <e, d>.

Fig. 3.20: A Petri net.

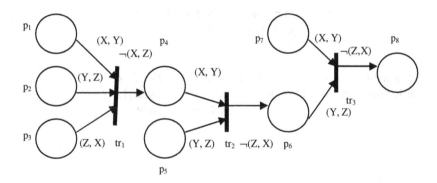

Token at the place p_1 = <a, b>, Token at the place p_2 = <b, c>, Token at the place
p_3 = <c, a>,
Token at the place p_5 = <c, d>, Token at the place p_7 = <e, d>, Token at the place
p_8 = ¬<a, e>.

Fig. 3.21: A Petri net

[**Hints:** The pipelining of transitions for the first and the second case are
given in Fig. 3.22 and Fig. 3.23 respectively.

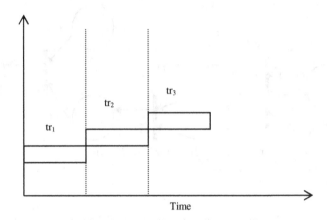

Fig. 3.22: Diagram showing pipelining of transitions for the first case

Here, in the case of the first Petri net, as per definition:

$$n = 8,$$
$$\Sigma s_i = s_1 + s_2 + s_3 = 4 + 2 + 2 = 8,$$
$$\Sigma m_i = \Sigma s_i + \Sigma d_i = 8 + 3 = 11,$$
$$k = 3$$

$\therefore$ Speed-up factor $S = T_u/T_m = n / (n - \Sigma s_i + k) = 8/3 = 2.66$.
& Resource utilization rate $\mu = \Sigma s_i / [k \cdot \max \{s_k : 1 \le k \le n\}]$
$$= 8/(3 \times 4) = 8/12 = 0.66.$$

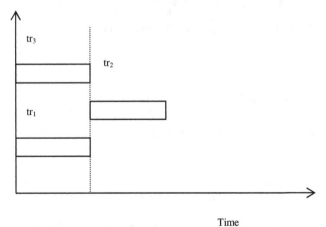

Fig. 3.23: Diagram showing pipelining of transitions for the second case

Now, for the second case,

$$n = 9,$$
$$\sum s_i = 4+2+3 = 9,$$
$$k = 2.$$

$\therefore$ The speed-up factor $S = T_u/T_m = n /(n - \sum s_i + k)$
$$= 9/(9 - \sum s_i + k) = 9/2 = 4 \cdot 5.$$
and Resource utilization rate $\mu = \sum s_i /[k \cdot max \{s_k: 1 \le k \le n\}]$
$$= 9/(2 \times 7) = 9/14 = 0 \cdot 64.]$$

31. The Petri net shown in Fig. 3.12 includes sets of concurrently resolvable clauses.

      (a) Using the proposed algorithm 'Procedure Automated-Reasoning', identify the transitions where concurrent resolutions take place in parallel.

      (b) Also show in the diagram the pipelining of transitions where concurrent resolution takes place.

      (c) Assume that the time required for concurrent resolution in a transition is proportional to the number of input and output concurrently resolvable clauses and hence determine the overall time of execution of the given logic program.

[Hints:

(a) If we use the proposed algorithm 'Procedure Automated-Reasoning', we can easily observe that in the first pass through the algorithm the transitions tr_1, tr_2 and tr_3 are concurrently resolvable as all those three transitions satisfy the three conditions required for firing according to the proposed algorithm. In the second pass, the transitions tr_1 and tr_2 are concurrently resolvable as can be easily found out. Then in the final third pass only the transition tr_3 resolves followed by the stopping condition of the algorithm.

(b) Figure 3.24 shows the pipelining of transitions where concurrent resolution takes place.

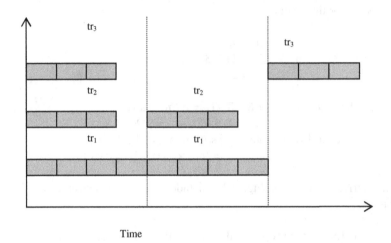

Time

Time to fire a transition which has been taken to be equal to the number of the clauses involved in the resolution process.

Fig. 3.24: Diagram showing pipelining of transitions for concurrent resolution

(c) Assuming that the time required for concurrent resolution in a transition is proportional to the number of input concurrently resolvable clauses, the overall time of execution of the given logic program becomes:

11t, where t = time taken for each input clause consideration for resolution.]

References

1. Bhattacharya, A., Konar, A. and Mandal, A. K., "A parallel and distributed computational approach to logic programming," proc. of *International Workshop on Distributed Computing (IWDC 2001)*, held in Calcutta, December 2001.
2. Bender, E. A., "Mathematical Methods in Artificial Intelligence," IEEE Computer Society Press, Los Alamitos, CA, 1996.
3. Ganguly, S., Silberschatz, A., Tsur, S., "Mapping datalog program execution to networks of processors," *IEEE Trans. on Knowledge and Data Engineering*, vol. 7, no. 3, June 1995.
4. Hermenegildo, M. and Tick, E.,"Memory performance of AND-parallel PROLOG on shared-memory architecture," *Proc. of the 1988 International*

Conference, on Parallel Processing, vol. II, Software, pp. 17-21, Aug. 15-19, 1988.

5. Hwang, K. and Briggs, F. A., *Computer Architecture and Parallel Processing*, McGraw-Hill, pp. 27-35, 1986.

6. Jeffrey, J., Lobo, J. and Murata, T., "A high–level Petri net for goal-directed semantics of Horn clause logic," *IEEE Trans. On Knowledge and Data Engineering*, vol. 8, no. 2, April 1996.

7. Kale, M. V., "Parallel Problem Solving," in *Parallel Algorithms for Machine Intelligence and Vision*, Kumar, V., Gopalakrishnan, P. S. and Kanal, L.N., (Eds.), Springer-Verlag, Heidelberg, 1990.

8. Konar, A., *Uncertainty Management in Expert Systems Using Fuzzy Petri Nets*, Ph.D. thesis, Jadavpur University, 1994.

9. Konar, A. and Mandal, A.K., "Uncertainty management in expert systems using fuzzy Petri nets," *IEEE Trans. on Knowledge and Data Engg.*, vol. 8., no.1, Feb. 1996.

10. Li, L., "High level Petri net model of logic program with negation," *IEEE Trans. on Knowledge and Data Engg.*, vol. 6, no. 3, June 1994.

11. Murata, T. and Yamaguchi, H., "A Petri net with negative tokens and its application to automated reasoning," *proc. of the 33rd Midwest Symp. on Circuits and Systems,* Calgary, Canada, Aug. 12-15,1990.

12. Naganuma, J., Ogura, T., Yamada, S-I., and Kimura, T., "High-speed CAM-based Architecture for a Prolog Machine (ASCA)," *IEEE Transactions on Computers*, vol. 37, no.11, November 1988.

13. Patt, Y. N., "Alternative Implementations of Prolog: the micro architecture perspectives," *IEEE Trans. on Systems, Man and Cybernetics.* vol. 19, no.4, July/August 1989.

14. Peterka, G. and Murata, T., "Proof procedure and answer extraction in Petri net model of logic programs," *IEEE Trans. on Software Engineering*, vol. 15, no. 2, Feb. 1989.

15. Takeuchi, A., *Parallel Logic Programming*, Wiley, 1992.

16. Yan, J. C., "Towards parallel knowledge processing," chapter 4, Advanced series on Artificial Intelligence, vol. 2, *Knowledge Engineering Shells: Systems and Techniques*, Bourbakis, N. G., (Ed.), World Scientific, Singapore, 1993.

4

Realization of a Parallel Architecture
for the Petri Net Model

The chapter provides hardwired design of a parallel computational engine for logic programming based on the reasoning algorithm outlined in chapter 3. The proposed engine takes care of the following two architectural considerations: (i) concurrent resolutions of multiple program clauses mapped onto a transition and its associated places, and (ii) concurrent resolution of several groups of program clauses distributed throughout the Petri net. Such concurrent resolution ensures AND-, OR-, Stream- and Unification-parallelisms of a logic program. The proposed architecture includes several stages of pipelines and thus supports massive parallelism among the modules within a pipelined stage. It also provides parallelism among the modules of different stages. A timing analysis of pipelining in the proposed architecture reveals that one complete transition firing cycle, that begins with a system reset and continues till writing of tokens onto an inert place, is approximately 100 T_c, where T_c denotes the time period of the system clock. A typical logic program that requires several transition firing cycles thus consumes a time proportional to 100 T_c. The constant of proportionality is fixed by the degree of parallelism of the proposed program and available system resources. The higher is the degree of parallelism, the lower is the proportional constant. Further, a reduction in system resources requires mapping part of the program onto same resources in real time, causing an increase in execution time.

4.1 Introduction

Chapter 1 introduced four ideal machines of parallel architecture depending on the instruction and data flow in a computer. The computational engines of these machines are popularly referred to as SISD, SIMD, MISD and MIMD architectures [1]. In this chapter we propose a new architecture for parallel processing, useful for logic programming applications. The architecture to be proposed shortly includes different control and data paths among its constituent modules, and thus it falls within the category of MIMD machines.

A typical logic program comprises of a set of clauses. Chapter 3 has demonstrated the scope of concurrent resolution of multiple program clauses on a

A. Bhattacharya et al.: *Realization of a Parallel Architecture for the Petri Net Model*, Studies in Computational Intelligence (SCI) **24**, 177–210 (2006)
www.springerlink.com

graphical engine like Petri nets. The current chapter provides an architectural framework for realization of the concurrent resolution of the program clauses. Since the framework is designed following the algorithm for automated reasoning presented in chapter 3, it ensures the concurrent execution of the four possible types of parallelisms in a logic program.

An examination of logic programs used in commercial database systems reveals that typical program clauses usually do not have more than five predicates/literals. Further, most applications employ binary predicates, i.e. predicates having two arguments. Taking into consideration of the degree of parallelism and utilization of hardwired resources, we restricted the number of predicates per clause = 5 and number of arguments per predicate = 2 in the proposed architecture. For practical reasons of resource limits, we consider at most two arcs a_1 and a_2 between a given place p_i and a transition tr_j, such that $a_1 \in (p_i \times tr_j)$ and $a_2 \in (tr_j \times p_i)$. Thus, multiplicity of parallel arcs directed similarly between a given place and a transition is not allowed. We further presumed that one place may be connected with at most two transitions. Extension of any of the above system resources, however, is permissible with minor changes in the system design.

Before closing this section, we briefly address some important and essential issues related to syntax. Logic programs, like any other programs, first needs to be compiled and the object codes may be run on a given machine. The source code in our system is a pseudo PROLOG program with all syntax identical, excluding the :- symbol, which we have intentionally replaced by ←, where A← B, C has a conventional meaning, as discussed in chapter 3. After the given program passes the compilation phase, a task allocation unit is employed to distribute and map the program clauses and data clauses onto different units of the architecture. The architecture is then ready to function. When the execution of the program is terminated, the control returns the response to the users. In this chapter, we, however, restrict our discussion to the architecture only.

The chapter is categorized into ten sections. Section 4.2 provides an overview to the overall architecture with special reference to six typical modules embedded therein. The detailed design of the individual modules is presented in Sections 4.3 through 4.8. The timing analysis of the proposed architecture is covered in Section 4.9. Conclusions are listed in Section 4.10.

4.2 The Modular Architecture of the Overall System

The proposed architecture consists of six major modules:

> (i) *Transition History File (THF) for transition tr_j, $1 \leq \forall j \leq n$;*
> (ii) *Place Token Variable Value Mapper (PTVVM) for place p_i,*
> *$1 \leq \forall i \leq m$;*
> (iii) *Matcher (M) for transition tr_j, $1 \leq \forall j \leq n$;*

(iv) First Pre-condition Synthesizer (FPS), realized with AND-OR Logic
for transition tr_j, $1 \leq \forall j \leq n$;
(v) Transition Status File (TSF) for transition tr_j, $1 \leq \forall j \leq n$;
(vi) Firing Criteria Testing Logic (FCTL) for transition tr_j, $1 \leq \forall j \leq n$.

Before execution of a logic program, a compiler, specially constructed for this purpose, is employed to *parse* the given program for syntax analysis. On successful parsing, the variables used in the programs are mapped onto a specialized hardwired unit, called *Transition History File (THF)* register. The compiler assigns the value of the variables, hereafter called *tokens*, at specialized hardwired units, called *Place Token Variable Value Mapper (PTVVM)*. The sign of the arc function variables is assigned to the PTVVM by the compiler. The compiler also constructs logic function for *First Pre-condition Synthesizer (FPS)* and assigns null value in the current and used binding fields of *Transition Status File (TSF)*, the details of which will be discussed in chapter 5. The functional behavior of the modules in the proposed architecture is outlined here.

On reset the THF for each transition activates the PTVVMs for the places associated with the given transition through appropriate *place-name* lines. Consequently, the activated PTVVMs address their internal place buffers for initiating the matching of arc function variables with the tokens already saved in the place buffers (places in Petri net terminology). These tokens are now compared with arc functions mapped at the arcs connected between one place and one transition (hereafter, referred to as adjacent place/ transition of an arc) to check consistency of variable bindings. The process of determining the consistency of the variable bindings of arc functions connected between an adjacent place/transition pair with the tokens residing at the said place is called *local matching*. Thus in case there exists two arc functions between a place and a transition, then the same initial token values of the adjacent place will be loaded into two place buffers (vide Fig. 4.4) for possible local matching of the two arc function variables. To maintain the identity of these two buffers, arc function tags are required. The THF generates these arc function tags through its control lines for subsequent labeling of the signed arc function variables in the place buffers. The detail of the labeling process, which is omitted here to avoid complexity, is covered in section 4.4.1. When there exists a single arc function between a place and a transition, the two arc function tags will have the same value; consequently the arc function variables will have an identical binding with respect to the tokens residing at the buffers inside the PTVVM.

Since a transition may have a number of associated places, the variable bindings generated for the arc functions connected between the given transition and its associated places need further to be compared to determine the consistent bindings (*most general unifier* [5] of the clauses participating in the resolution process). This is referred to as *global matching*. For global matching of the arc function variables, say X, Y, Z, a *matcher* circuit M is employed with each transition. This circuit receives the local variable bindings of the arc functions, associated with all the arcs connected with the transition from the PTVVMs, and then determines the global consistency of the variables present in the arc functions.

The PTVVM for each place associated with a transition generates a flag signal, indicating existence of tokens at that place. The FPS circuit for each transition grabs these flags from the PTVVMs to determine whether all excluding at most one place contains tokens of same arity as that of the respective arc functions. In our implementation, we consider all arc functions comprising of two variables, and thus we need to check only the presence of tokens at all places, leaving at most one, and need not bother on arity matching.

On reset, the TSF for tr_j checks whether the current set of binding received from the matcher M for tr_j is a member of the used (cumulative) set of binding for the same transition. It also issues a single bit flag signal to represent the status of the condition referred to above. The *Firing Criteria Testing Logic (FCTL)* circuit is employed for each tr_j to test the joint occurrence of the three preconditions received from FPS Logic, the matcher M and the TSF. If all these three conditions are jointly satisfied, the FCTL issues a *'fire tr_j'* command, indicating the right time of firing the transition.

The PTVVM of the inert place associated with the fired transition on receiving this signal saves the new (signed) token at its internal buffer.   The value of the consistent current-bindings set generated at the matcher M is sent to the TSF of the corresponding tr_j. The process is continued until no new consistent bindings are generated. It may be added here that a number of transitions that jointly satisfy all the three pre-conditions, fire concurrently (vide chapter 3) and the complete control of token generation and their placement at the inert places is taken care of by the proposed architecture.

4.3 Transition History File

The Transition History File (THF) keeps track of the histories for all transitions in the Petri net. It is realized with a register file comprising of n registers, where the register r_j, $1 \le j \le n$, contains the history of transition tr_j. For instance, the THF in Fig. 4.2 has two registers corresponding to the transitions tr_1 and tr_2 in the Petri net of Fig. 3.7 (vide chapter 3). Each register comprises of three distinct types of fields namely (i) transition name tr_j, (ii) place p_i associated with the transition denoted by APp_i and (iii) arc function associated with p_i denoted by $AFAWp_i$.

For simplicity and convenience, the fields of the registers in Fig. 4.2 have been designated with reference to the Petri net in Fig. 3.7. Since tr_1 in Fig. 3.7 has three associated places p_1, p_2, p_3, the respective APp_k, APp_l and APp_m fields contain p_1, p_2 and p_3 respectively. In general, we consider at most five places associated with the transition, the last two been APp_r and APp_o respectively. In the present context

APp$_r$ and APp$_o$ assume null values. These places are not shown in the figure for neatness and clarity of presentation. These additional fields should be assigned with null values.

The arc functions corresponding to the five places called AFAWp$_k$, AFAWp$_l$, AFAWp$_m$, AFAWp$_r$, AFAWp$_o$ presume values A, B, C, $\varnothing$ and $\varnothing$ respectively. The details of A, B and C are presented in the label of Fig. 4.2 itself. The control lines that carry the signals for the signed arc functions, such as +XY¬YZ for the label A in Fig. 4.2 corresponds to the two arc functions associated with place p$_1$ and transition tr$_1$ in Petri net of Fig. 3.7. The sub-field definitions of label A must be ordered to ensure the variables in an arc function occupy the next immediate position of its sign bit, as shown in the last example. The other arc functions AFAWp$_r$ have similarly been constructed and shown in Fig. 4.2 accordingly.

When a logic HIGH level appears at the input of all the registers, the associated place fields (APp$_i$), having non-null values, yield a HIGH logic level for activation of the appropriate PTVVM for a place. Thus when p$_1$ line from register for tr$_1$ in Fig. 4.2 is HIGH, the PTVVM for place p$_1$ is activated.

The control signals AFAWp$_i$ are also generated concurrently with the place fields APp$_i$ just after power-on. The AFAWp$_i$ fields are thus transferred as the control signals to the PTVVM for subsequent actions.

4.4 The PTVVM

The PTVVM comprises of three main sub-units (vide Fig.4.3). The first sub-unit holds the data in its internal buffers to initiate the local token matching. It also controls the mode selection to initiate either of the following two alternatives: i) local token matching and ii) generation of signed token for the inert place. The second sub-unit performs the local token matching, while the third sub-unit determines the existence of tokens in the concerned place, and submits the status to the FPS by a flag. The design of these modules is presented as follows.

4.4.1 The First Sub-Unit of the PTVVM

While designing the architecture, two arcs between a place and a transition is presumed to limit system resources. Since the variables in the arc function of these two arcs might have different bindings, two place buffers are needed to realize this, which is depicted in the first module of the PTVVM in Fig. 4.4. It is assumed here that there is a provision for a maximum of five tokens in a place, accordingly, the place buffers (vide grid-like units in Fig. 4.4) are designed to have five accessible locations. Further, for limiting the width of a signed token to a maximum of three, the words of the place buffers are presumed to have three fields. The first field of the token holds the sign of the token, while the last two fields stand for the constant bindings of the variables in the arc function. When there exists only one arc between a place and a transition, the contents of the two place buffers are made identical for the generalization of the system design. It may further be noted that arc function tags are also recorded in separate registers.

The arc function tags (AFT1 and AFT2) shown in the Fig. 4.4 correspond to the signed arc functions associated with the arcs connected between a place and a transition. For instance, the arc function tags AFT1 and AFT2 hold $+ X Y$ and $\neg Y Z$ (not shown for lack of space) respectively corresponding to the arc functions for the arcs connected between transition tr_1 and place p_1 of Fig. 3.7 (vide chapter 3). It may be emphasized here again that the signed arc functions are received from the THF through control lines after power-on. Since the design allows a connectivity of at most two transitions with a place, two more arc function tags for the second transition are needed. These arc function tags are not shown in the figure to avoid clumsiness in the drawing.

The following three major tasks are required to be performed with the help of the place buffers. First, the place buffers are needed to hold the initial tokens like the markings recorded in a place. Secondly, the place buffers play an important role in local token matching, which, however, is executed in the second sub-unit of the PTVVM. Lastly, consequent to firing of a transition, the resulting new tokens may be saved in the place buffers of the place associated with the transition. The control logic circuit, shown at the right bottom part of Fig. 4.4, generates the necessary control commands to execute the last two tasks.

For enabling the local token matching to be executed in the second sub-unit of the PTVVM, a mode selector logic (MSL) (vide bottom part of Fig. 4.4) resets the flip flop FF, which subsequently activates both the switches S_2 (located near FF) and S_2' (located at left top corner)· On closure of the switch S_2, the synchronous clock source (SCS) activated by the logic HIGH signal AP_X received from the THF resets counter C3 (located close to the OR-gate) to initiate counting. The switch S_2' being closed, the counter C2 also gets reset for subsequent counting. It needs mention that the clock rate for the counter C2 is $1/5^{th}$ of the clock rate for the counter C3. The counters C2 and C3 are used for generating the address of the place buffers PB_1 and PB_2 respectively. The speed of the counter C2 being $1/5^{th}$ of the speed of C3, the contents of all the five locations of PB_2 are compared with each content of PB_1. This is needed to test the consistency in local token matching.

To store a new value in the place buffer, the MSL in bottom of Fig. 4.4 presets the flip flop FF for subsequent comparison of the content of the matched value register (MVR) (located at top of Fig. 4.4) with the content of place buffer PB_2. On presetting, the flip flop FF closes the switch S_1 to initiate counting at the counter C3. The content of the MVR, assuming switch S_3 is closed, is compared with the token of the place buffer PB_2 from the location addressed by the counter C3. A control circuit, not shown in Fig. 4.4, is needed to match the respective fields of the MVR and the PB_2. The objective of the comparison is to determine whether the content of the MVR already resides in place buffer PB_2. If the result is in the negative, the contents of the MVR needs to be saved in one blank location within the place buffer PB_2. This has been realized here by generating a control command (zz) at the output of the comparator. A value of zz = 1 indicates that the content of the MVR is different from that of the current value of place buffer PB_2. This control command enables a null register R2 to output a null value designated by a 4-bit data-stream 1111 for comparison with the current content of the place buffer PB_2 via register R1 with the help of the null checker circuit NC.

If R1 (beside TSB) contains a null value 1111, the NC issues a HIGH control command to close the switch S_4 (near TSB), which establishes a data path from the MVR to the register D in the second sub-unit of the PTVVM, provided S_3 is still closed. If at least one transition associated with the place of the corresponding PTVVM fires, the switch S_3 gets closed. In order to prevent the scrambling of new token values entering a common place through multiple firable transitions, AND gates are used to steer the value of the variables to the Local Token Matcher (LTM) circuit of the second sub-unit for the respective firable transitions.

The counter C3 is also used in the third sub-unit of the PTVVM for flag generation signifying existence of tokens at a place.

4.4.2 The Second Sub-Unit of the PTVVM

The second sub-unit of the PTVVM (vide Fig. 4.5) comprises of separate Local Token Matcher circuits (LTM) for different transitions associated with the place corresponding to the PTVVM (vide Fig. 4.6). The signs of the arc functions are stored in the arc function registers AF1 and AF2 (vide Fig. 4.6) in the respective fields of the arc functions A and C of the LTM in the second sub-unit. Using the control lines 1 and 2, the signed tokens are stored in the appropriate fields (Sign, X, Y and Z) of the registers B and D associated with AF1 and AF2 respectively in accordance with the tags such as AFT1 and AFT2 of the place buffers (vide Fig. 4.4). The sign comparators together with the value comparators followed by the AND logic determine the consistent bindings of X, Y, Z which are transferred to the matcher circuit. For a common place p_i connected to a number of transitions the appropriate hardware are replicated.

A little thought will reveal that only one out of all the places associated with a transition, will be prevented from generating variable bindings of the arc function. The arc function unsuccessful in the variable binding process is called the inactive arc function. A token while entering a place takes the opposite sign of the inactive arc function that steers it to the place. Sign complementers and additional registers R3 and R4 are employed in Fig. 4.6 to ensure this during the traversal of the signed tokens to the place buffers. The tri-state buffers, incorporated in the circuit, have their generic use of prohibiting signal flow in undesirable directions.

When an arc-function-register, AF contains a null value or no appropriately matched (signed) token, a control command is issued for transferring the properly signed token to the place buffers following the inactive arc function. The control logic is governed by two issues: first it checks whether the incoming token-stream at the arc function register is null (1111) and whether there already exist a matching token in the place buffer corresponding to the arc function. If the former condition is found true and the latter is false, then the signed token held by the additional registers R3 or R4 are passed on to the place buffers. The control logic circuit includes a null checker (NC) and a few logic gates to identify the inactive arc function in each PTVVM module. It also allows the relevant bus lines to transmit the signed tokens to the place buffers. Otherwise the transmission of the signed tokens is aborted.

The tokens held by the place buffers are transferred to the third sub-unit one by one as addressed by the first sub-unit for subsequent checking of the existence of at least one token at the place. The existence of tokens at the places associated with a transition is required for testing one pre-condition of firing of the transition.

4.4.3 The Third Sub-Unit of the PTVVM

The status Flag Generating Circuits (FGC), incorporated in the third sub-unit of the PTVVM (vide Fig. 4.7), checks if there exists at least one non-null token in place p_i. This circuit receives the content of all the locations of place buffer PB_2 (copy of which is retained in place buffer PB_1) with the help of a counter (vide Fig. 4.4) and a decoder. The decoder activates the switches S_6 to S_{10} in sequence, so as to get the proper contents from each locations of the place buffer PB_2 in corresponding registers of FGC. The AND gates connected with this register checks whether the content of the location is null (1111). Thus a ZERO in the output of any one AND gate will propagate as a ZERO through next level AND gates making the output status line of FGC to be ONE indicating that at least one of the tokens in place p_i is non-null.

4.5 The Matcher

The matcher circuit M receives signal for a single variable, say X, from different places associated with a given transition tr_j. Since the value of the variable X could assume either a constant or null bindings, an arrangement has been made to ensure consistency among these possible bindings. For instance, if the value of a variable X obtained from one place associated with the transition is "a" and the value of X obtained from the remaining places connected with the transition is "null"(denoted by 1111), then the resulting binding for X is "a". However, if value of X obtained from the places associated with a transition is "a", "b" and "null values" respectively, no consistent bindings can be formed. On evaluation of a consistent binding for a variable, the Matcher circuit (vide Fig. 4.8) generates a logic level HIGH signal at its output for subsequent activation of the FCTL circuit. This is realized in the Matcher circuit of Fig. 4.8 in two phases.

In the first phase the value of variable X, denoted by 4-bit strings, are received from the PTVVM of the respective places at the Xp_1, Xp_2,...,Xp_5 lines of the

Matcher (Fig. 4.8). These values are stored in the internal registers R1, R2,..., R5 of the Matcher respectively. The Matcher in the first phase checks whether the contents of the registers are null. If yes, the AND gates below the matcher yields a logic value HIGH, else it is ZERO. The OR gates above the MOS switches control the closure/opening of the switches based on the output of the AND gates below the registers. If a register contains a null value (i.e., 1111), and its right neighbor contains a constant (non-null) value (other than 1111), then the OR gate activates the MOS switches to short-circuit their drain-source terminals, thereby providing a wired-ANDing [3] of 4-bit contents of a register with its right neighbor. In case two successive registers contain different non-null values, the OR-gates' both input being ZERO, it remains inactive, and thus the MOS switches remain open.

The Matcher circuit in the second phase is engaged to compare the contents of each two successive registers. This is realized by using 4-bit comparators employing XOR gates. Since similar inputs of a XOR gate results in ZERO at its output, we need to invert them and AND the resulting signals to describe the results of comparison of two registers' contents by a single bit. An AND tree employed in Fig. 4.8 finally determines whether a variable X has any consistent bindings.

It may be added here that one of the pre-conditions for firing a transition is to determine a consistent binding of all variables associated with its arc functions. Assuming that arc functions contain three variables: X, Y, and Z, we need to determine the consistent bindings for all the three variables X, Y and Z. A transition can fire only after determining a consistent binding of all the three variables associated with its arc functions. This has been realized in the circuit diagram presented in Fig. 4.9 by employing three Matcher circuits for the three variables, and by one AND gate. The AND gate checks whether the outputs of each matcher resulted in a consistent bindings. Thus a HIGH level at the output of the AND gate satisfies one pre-condition for firing. The other AND gate shown in Fig. 4.9 ensures satisfaction of all the three pre-conditions of firing (see chapter 3 for details) a transition tr_j. The latter AND gate is referred to as the Firing Criteria Testing Logic (FCTL).

4.6 The Transition Status File

On reset, the Transition Status File (TSF) for each transition begins checking whether the set of current binding is a member of the set of used binding. In Fig. 4.10 we kept provision for four such X-Y-Z triplet fields that hold the used bindings and one additional triplet to represent the current binding. To test the existence of the current triplet in the used four triplets, a counter (C), an address decoder and a multiplexure (MUX) are needed. The MUX on receiving an address i, $0 \leq i \leq 3$, from the counter transfers the i-th triplet of the used binding to the three step comparator (3SC). The 3SC compares the respective X-, Y- and Z-fields of the current and used binding space. In case the current set is not a member of the used bindings, then the current set should be placed in an empty slot of the used bindings. The empty slots actually contain a null value (1111) and thus we can easily check its status by employing a null checker NC2 that works in parallel with the 3SC. It is clear from the Fig. 4.10 that both NC2 and the 3SC receive common input from the MUX. Now, suppose at count = j, the current set is detected not to be a member of the used bindings, and the j-th slot has been found empty (denoted by a null value). Consequently, the counter remains held up at count = j, and the j-th slot of the set of used binding is filled in with the current value.

It may be noted that initially the current triplet and the used four triplets are all initialized to be null (1111) by the compiler. The circuit thus starts functioning with null values, until a new value is received at the current triplet field from the matcher M. The fields of current binding are subsequently updated by the respective contents of the matcher M.

A flag is used to indicate two special circuit conditions: (i) whether the set of current binding is non-null and (ii) whether the set of current binding is available in one of the fields of the used binding. When any of the aforementioned conditions are found to be true, the flag is HIGH, else it remains LOW. Thus in case all transitions' current binding are members of a corresponding set of used binding, the flag signal for each transition will be 1. An AND-gate (vide Fig. 4.11) is used to test the joint occurrence of all the HIGH flags. The resulting output of the AND-gate needs to be inverted to pass it on to a Low logic level message to the FCTL, signifying that all transitions' current bindings are members of their used bindings, and consequently firing of transitions are no longer needed.

4.7 The First Pre-condition Synthesizer

One important prerequisite for firing a transition is the existence of tokens at all places barring at most one. This has subsequently been referred to as the first pre-condition for firing a transition. To illustrate this principle, let us consider a transition having a total of five input and/or output places. Let a, b, c, d and e be five Boolean variables designating presence/ absence of tokens at those five places. Thus the condition that all except at most one place possess tokens can be described by Boolean expression (4.1), where the superscript "c" over a variable denotes its complementation.

$$abcde + a^c bcde + ab^c cde + abc^c de + abcd^c e + abcde^c \qquad (4.1)$$

The expression (4.1) can be simplified easily to the following form by ORing abcde to each term of (4.1) starting from the second term.

$$bcde + acde + abde + abce + abcd. \qquad (4.2)$$

The Petri net we used in Fig. 3.7, however, has three places associated with each transition. Thus expression (4.1) reduces to a three-variable Boolean function, described by

$$abc + a^c bc + ab^c c + abc^c \tag{4.3}$$

which can be simplified to

$$ab + ac + bc. \tag{4.4}$$

The Boolean variables in the present context are flags generated by the PTVVM, indicating whether the place contains any tokens. For convenience in understanding, we denote the flags generated by a place by the place name itself. Thus the Boolean function for tr_1 and tr_2 (vide Fig. 3.8) are given by

$$p_1p_2 + p_2p_3 + p_3p_1 \tag{4.5}$$
$$p_3p_4 + p_4p_5 + p_5p_3 \tag{4.6}$$

Figure 4.12 presents a logic circuit for the above two Boolean expressions. The circuit receives flags from the respective PTVVMs of places p_1, p_2, p_3, p_4 and p_5 respectively, and communicates the status of presence/ absence of tokens at all excluding one input/output places of a transition to the Firing Criteria Testing Logic (FCTL). It is to be noted that the FPS includes the logic circuitry corresponding to all the transitions in the Petri net.

4.8 The Firing Criteria Testing Logic

The Firing Criteria Testing Logic (FCTL), vide Fig. 4.9, is needed to decide the possible firing of a transition based on the joint occurrence of the following three criteria.

(1)   All excepting at most one place associated with the transition possess tokens;

(2)   All the variables in the arc functions associated with a transition yields a globally consistent value; and

(3)   The current binding is not a member of the used binding or the current binding is a null vector.

For each of the aforementioned three conditions one flag is generated, the joint occurrence of which is tested by the FCTL. The circuit is simple. It just includes a 3-input AND gate for each transition. The output of the AND gate carries the current status of firing of the corresponding transition. The status signal thus generated is carried to the PTVVM for requisite subsequent actions.

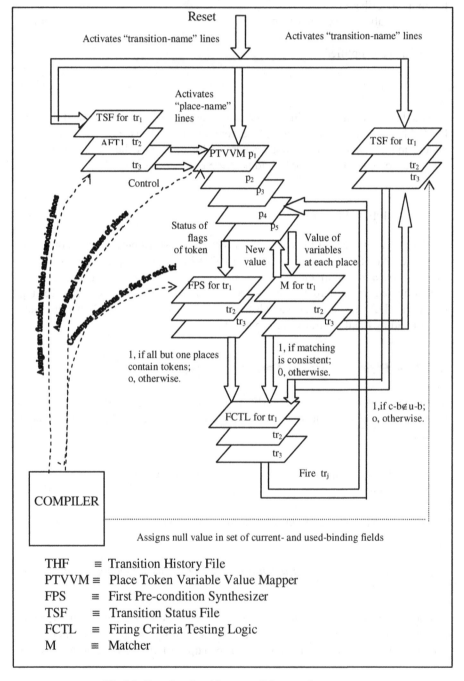

THF	≡ Transition History File
PTVVM	≡ Place Token Variable Value Mapper
FPS	≡ First Pre-condition Synthesizer
TSF	≡ Transition Status File
FCTL	≡ Firing Criteria Testing Logic
M	≡ Matcher

Fig 4.1: Functional architecture of the complete system

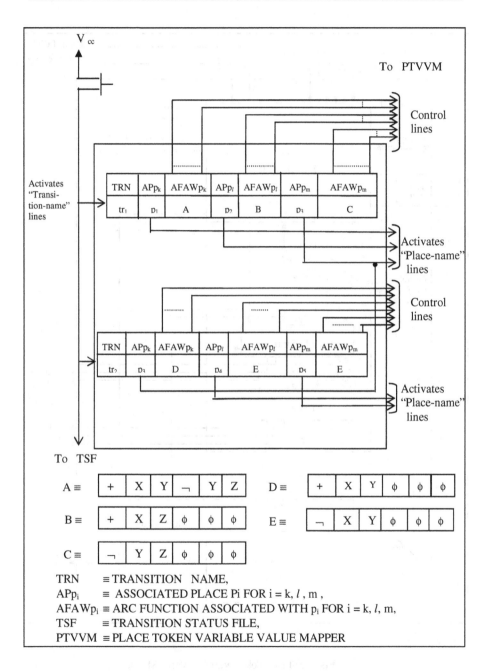

TRN   ≡ TRANSITION   NAME,
APp_i ≡ ASSOCIATED PLACE Pi FOR $i = k, l, m$,
$AFAWp_i$ ≡ ARC FUNCTION ASSOCIATED WITH p_i FOR $i = k, l, m$,
TSF   ≡ TRANSITION STATUS FILE,
PTVVM ≡ PLACE TOKEN VARIABLE VALUE MAPPER

Fig 4.2: Transition History File

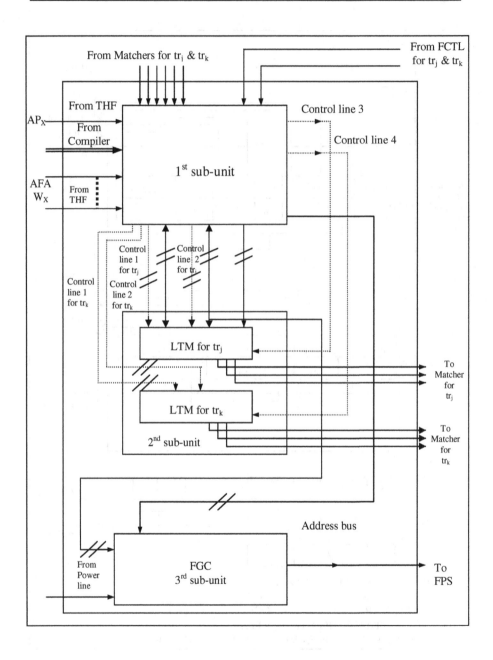

Fig. 4.3: Place Token Variable Value Mapper

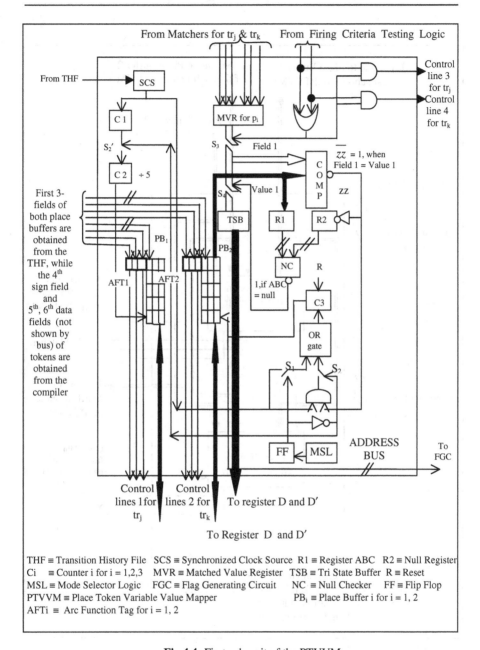

From Matchers for tr_j & tr_k From Firing Criteria Testing Logic

Control line 3 for tr_j
Control line 4 for tr_k

From THF

SCS

C 1

S_2'

C 2   ÷ 5

First 3-fields of both place buffers are obtained from the THF, while the 4th sign field and 5th, 6th data fields (not shown by bus) of tokens are obtained from the compiler

MVR for p_i

S_3 Field 1

$\overline{ZZ}$ = 1, when Field 1 = Value 1

C O M P   ZZ

S_4 Value 1

TSB   R1   R2

1,if ABC = null

NC   R

PB_1 PB_2

AFT1   AFT2

C3

OR gate

S_1 S_2

FF   MSL

ADDRESS BUS   To FGC

Control lines 1 for tr_j Control lines 2 for tr_k To register D and D′

To Register D and D′

THF ≡ Transition History File   SCS ≡ Synchronized Clock Source   R1 ≡ Register ABC   R2 ≡ Null Register
C_i ≡ Counter i for i = 1,2,3   MVR ≡ Matched Value Register   TSB ≡ Tri State Buffer   R ≡ Reset
MSL ≡ Mode Selector Logic   FGC ≡ Flag Generating Circuit   NC ≡ Null Checker   FF ≡ Flip Flop
PTVVM ≡ Place Token Variable Value Mapper PB_i ≡ Place Buffer i for i = 1, 2
AFTi ≡ Arc Function Tag for i = 1, 2

Fig 4.4: First sub-unit of the PTVVM

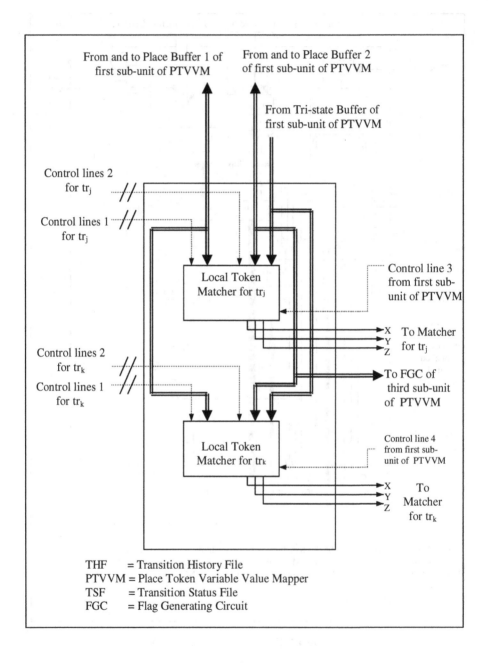

Fig. 4.5: Second Subunit of PTVVM for p_i

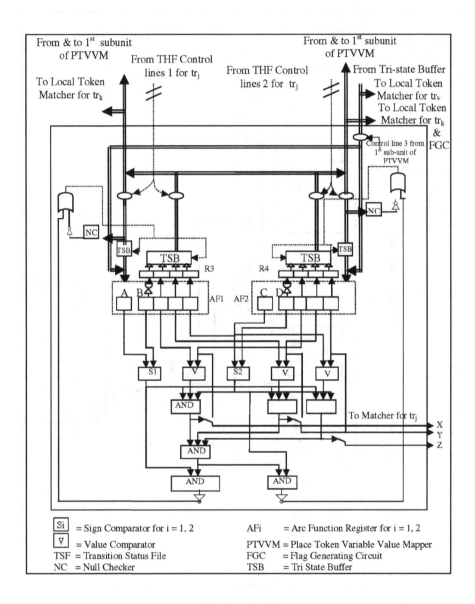

Fig 4.6: Local Token Matcher for tr_j in 2nd subunit of PTVVM for p_i

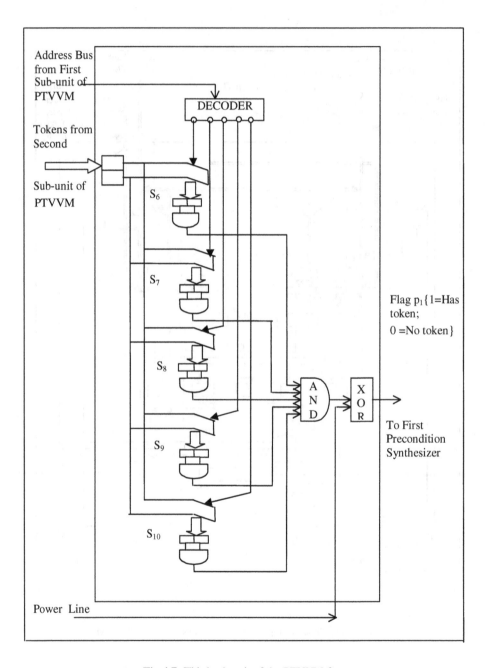

Fig 4.7: Third sub-unit of the PTVVM for p_i

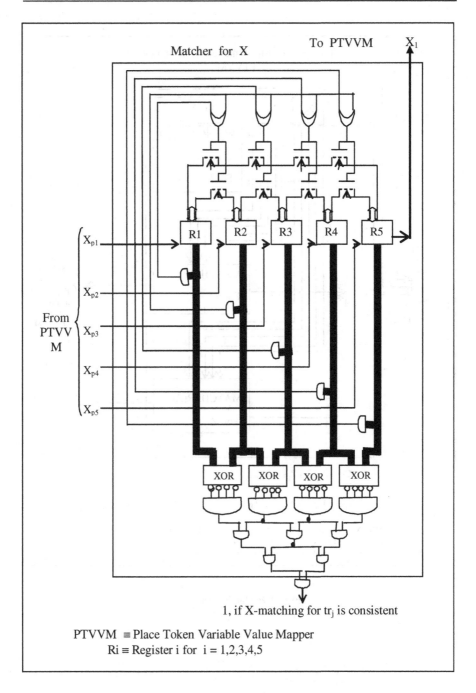

Fig. 4.8: Matcher for variable X for tr_j

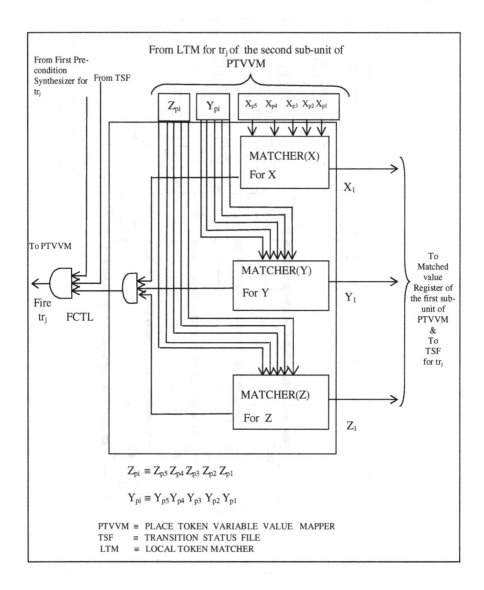

Fig. 4.9: Matcher for tr_j

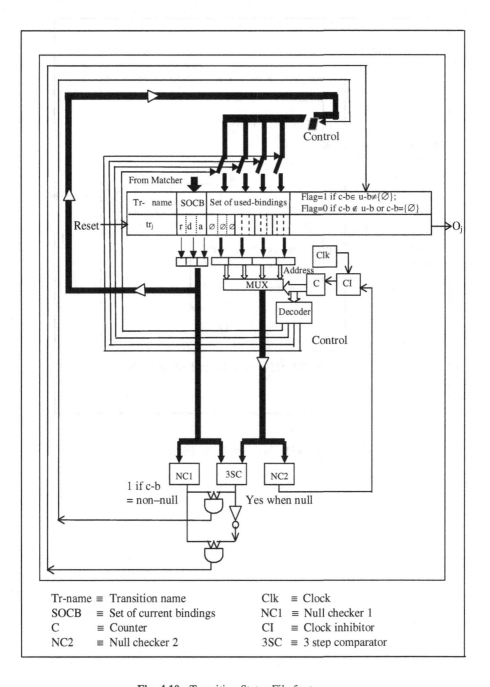

Fig. 4.10: Transition Status File for tr$_j$

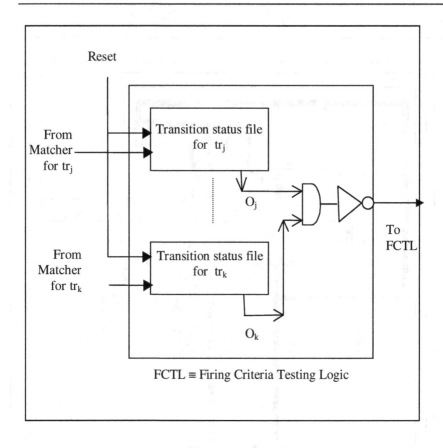

Fig. 4.11: Transition Status File

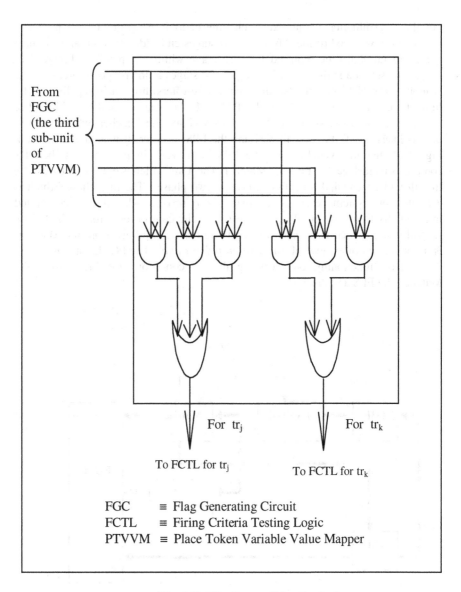

Fig. 4.12: First Pre-condition Synthesizer

4.9 Timing Analysis for the Proposed Architecture

For the determination of the execution time of logic programs on the proposed architecture, we need to identify pipelined stages embedded therein, and estimate the computational time required for each stage within the pipeline. Figure 4.13 provides a schematic diagram of the pipelined stages in the proposed architecture. A look at Fig. 4.13 reveals that the architecture offers three main pipelines. The first one comprises of the THF, the PTVVM, the Matcher and the TSF. The second one comprises of the THF, the PTVVM and the Matcher, while the third one includes the THF, the PTVVM and the FPS in order. It is also evident from Fig. 4.14 that the Matcher and the FPS both work in parallel, but the FPS completes its task earlier than the Matcher. The TSF, which is employed to test the firability of an enabled transition, works in two phases. The first phase, which is activated on system reset, is a *waste phase (cycle)* [4]. It is incorporated intentionally to avoid complexity in designing the control logic for the TSF. The second phase, which is initiated on receiving current-bindings from the Matcher, is an effective component of a *transition firing cycle*[1]. The FCTL, shown in Fig. 4.13, starts functioning on receiving the pre-conditions for firing from the Matcher, the FPS and the TSF.

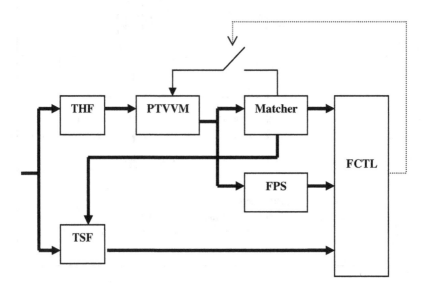

Fig. 4.13: Pipelining among the major modules of the architecture for the execution of FOL programs. The solid bold line denotes data lines, the solid thin line denotes data line for transmission of data for the next cycle, and the dotted line denotes 'fire transition' control signal

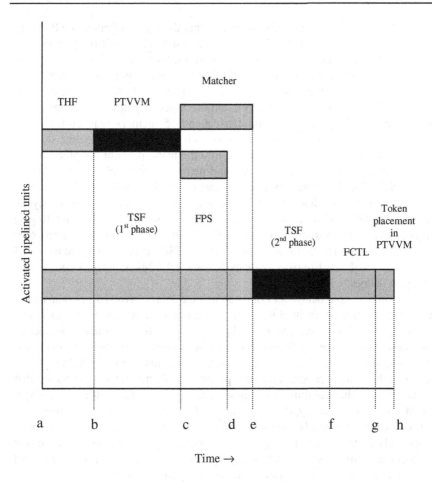

Fig. 4.14: Gantt chart showing concurrent and pipelined activation of different units in one transition firing cycle for evaluation of the cycle time

A timing analysis of the proposed architecture reveals that maximum time of a *transition firing cycle* is consumed by the PTVVM. An insight view of Fig. 4.3 and 4.5 further reveals that local matching of tokens in a PTVVM is primarily accomplished by determining the common variable bindings of two arc functions associated with one place and a transition. In our architectural realization, we presumed that a place contains at most five tokens. Thus, the variables in an arc function can have at most five bindings. Consequently, to determine the common variable bindings of two arc functions connected between a place and a transition,

---

[1] A transition firing cycle is defined as the interval of time between the issue of a system reset signal to placement of tokens at an inert place after firing of a transition.

we need $(5 \times 5) = 25$ comparisons. Assuming that the place buffers in the PTVVM (Fig. 4.4) are realized with RAMs driven by a system clock of time period T_c, it can easily be ascertained that 25 *memory read cycles* [2] (or 75 T_c) are needed to complete the local matching of tokens. Further, for storing an inferred token into a place buffer (Fig. 4.4) we need to compare the content of *Matched Value Register* (MVR) with the content of one place buffer. Since a place buffer contains 5 tokens, we need 5 comparisons and consequently 5 memory read cycles in the worst case for matching. One *memory write cycle* is also employed (time slot g to h in Fig. 4.14) to enter the inferred token into the blank place of the place buffer. The time consumed for 5 memory read and memory write cycles is $15 + 4 = 19$ clock cycles. Thus PTVVM approximately requires $(75 + 19)\ T_c = 94\ T_c$.

Among the other modules in Fig. 3.13, the THF requires 1 register access cycle, and the FPS, the FCTL and the Matcher consume approximately 2, 1 and 10 gate delays respectively. Assuming that the TSF circuit (Fig. 4.10) is realized with register files, a simple analysis shows that the TSF approximately consumes 4 clock cycles (or 4 T_c), 1 register write cycle (or 1 T_c), 1 MUX-delay, 1 comparison and 3 gate delays.

Thus ignoring gate delays and other switching delays involved in comparison, the time incurred in a transition firing cycle, starting from reset to token placement at a place is approximately $(94 + 4 + 1)T_c = 99\ T_c \approx 100\ T_c$. Since a number of transitions are concurrently firable, it is expected that the execution of a complete logic program will require an integer multiple of this transition firing cycles. Assuming that a logic program requires p transition firing cycles, the time needed for execution of the program thus is approximately 100 p Tc. Further, assuming a 1000M-Hz clock frequency, the time required for execution of a logic program on the proposed architecture is $(100 \times p \times 10^{-9}) = 0.1$ p μ-sec (microsecond). To have an idea about the execution time of commercial programs on the proposed architecture, let us consider a database program containing 2000 rule clauses and 8000 data clauses. Such programs can be configured on a Petri net with 2000 transitions[3] and approximately 2000 places[4], each place been mapped with 4 data clauses. Since the worst case value of p cannot exceed the number of transitions $(= 2000)$, we assume p = 2000. Consequently, time required for execution of the program on the proposed architecture = $0.1 \times 2000\ \mu$-sec = 200 μ-sec only.

4.10 Conclusions

This chapter presented a parallel architecture for logic programming based on the reasoning formalisms of Petri net discussed in the last chapter. The proposed architecture supports concurrent resolution of multiple program clauses associated

---

[2] Memory read and write cycle require 3 and 4 clocks respectively.

[3] No. of transitions = No. of rule clauses.

[4] Usually number of places in a Petri net is approximately equal to the number of transitions [2].

with each transition. Thus groups of concurrently resolvable clauses mapped at transitions and their adjacent input/output places can be resolved concurrently, thereby increasing the throughput to a great extent. An analysis of the proposed architecture reveals that there exist two multi-stage pipelines and parallelism among the modules of the architecture. Consequently with number of program clauses equal to the number of transitions, the architecture achieves a high degree of parallelism. The Gantt chart [6] shown in section 4.9 (vide Fig. 4.14) reveals that the time required for execution of a logic program on the proposed architecture is approximately 100 p T_c, where p and T_c denote the number of concurrent transition firing cycles and time period of the system clock respectively. Since p usually is of the order of 1000s, and T_c is of the order of microseconds, the resulting time is only of the order of one tenth of a second. Thus the proposed architecture will find massive applications in the database systems realized with Datalog programs.

Exercises

1. Given a Petri net with necessary labels of arc functions and associated places of transitions. Construct the Transition History Files. You need not show the input and output lines of the register files.

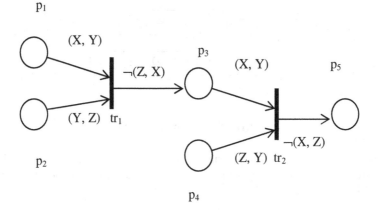

Fig. 4.15: A given Petri net

[**Hints:** The Transition History File for the Petri net given in Fig. 4.15 is constructed in the next page vide Fig. 4.16.

Transition	APp_k	$AFAWp_k$	APp_l	$AFAWp_l$	APp_m	$AFAWp_m$
tr_1	p_1	A	p_2	B	p_3	C
tr_2	p_3	D	p_4	E	p_5	F

Where,

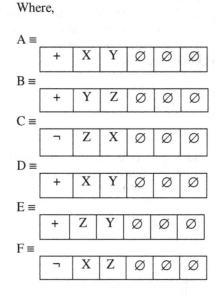

$A \equiv$

| + | X | Y | ∅ | ∅ | ∅ |

$B \equiv$

| + | Y | Z | ∅ | ∅ | ∅ |

$C \equiv$

| ¬ | Z | X | ∅ | ∅ | ∅ |

$D \equiv$

| + | X | Y | ∅ | ∅ | ∅ |

$E \equiv$

| + | Z | Y | ∅ | ∅ | ∅ |

$F \equiv$

| ¬ | X | Z | ∅ | ∅ | ∅ |

Fig. 4.16: Transition History File]

2. Suppose the place/transition names and arc functions are coded as binary strings of 3 and 12 bits respectively in a **Transition History File** register (THF).

   (a) Determine the word-length of the **Transition History File** registers.

   (b) Using flip-flops as the basic elements, design a complete register to realize the THF.

[Hints:

(a) Word-length of transition history file register = Word-length for one transition + Word-length for 3 place-names + Word-length for 3 associated arc functions = $(3 + 3 \times 3 + 3 \times 12)$ bits = 48 bits.

(b) To design the THF register the following points may be taken into account.

  (i) All flip-flops in each register should be activated simultaneously by the same clock.
  (ii) Except transition-name, other information of the THF need to be transferred to other units of the circuit, all in the same time.]

3. Suppose the predicates in a logic program include only three variables X, Y and Z. Assuming a '+' signed arc function corresponds to the input arc function and a '¬' signed arc function corresponds to the output function of a transition, determine the word-length required to represent the '$AFAWp_i$' in THF. Note that an arc function contains only two variables out of X, Y and Z.

[Hints: $AFAWp_i$ includes 2 arc functions, the format of which is given below.

+	X	Y	¬	Y	Z

where (X, Y) and (Y, Z) denote the input and the output arc functions respectively of the given transition.

To represent three variables we, minimally require two bits. Thus, let X, Y and Z be denoted by 00, 01 and 10 respectively. The '+' and the '¬' and 'don't care' sign requires two bits, say, '00' for '+', '11' for '¬' and '01' for 'don't care'. Thus the word-length becomes

+	X	Y	¬	Y	Z

Bit requirement: $2 + 2 + 2 + 2 + 2 + 2$

Thus, summing the bit requirements for individual bit we get the result to be 12 bits.

The individual numbers in the last expression corresponds to the bit requirement for the respective element in the given string.

It is to be noted that a null variable and sign can be denoted by "11" and '01' respectively.]

4. Consider the Petri net shown in Fig. 3.7. Given the time required for the following operations:

Operations	Time required for the operation
Local Token matching	30 μS
Testing the first and the third precondition	20 μS
Global token matching at the Matcher unit	10 μS

Determine the overall time required for firing transition tr_1.

[**Hints:** The order of execution at the modules inside the PTVVM given in the timing diagram (Fig. 4.17). It is clear from the diagram that the total time consumed = (30 + 10 + 20) μS = 60 μS.

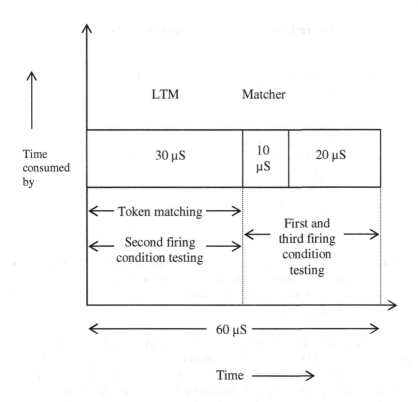

Fig. 4.17: Timing diagram describing order of execution at the modules
inside PTVVM for the Petri Net of Fig. 3.7]

5. Given the following timings in connection with transition firing in a Petri net.
   Assuming that the LTMs and Matchers of transitions tr_1 and tr_2 work in
   parallel, determine the total time needed for firing of both the transitions in
   Fig. 3.7.

Operations	Time required for the operation
Matching a single variable in an arc function	5 μS
Testing the first and the third precondition for firing	20 μS
Global token matching at the Matcher unit	10 μS

[**Hints:** The local token matching of the arc functions (X, Y), (X, Z) and ¬(Y, Z) can be done concurrently and the time needed for doing so

$$= (5 \ \mu S/variable) \times 2 \text{ variables}$$
$$= 10 \ \mu S.$$

The variable bindings thus obtained are transmitted to the Matcher for global token matching. It is indeed important to note that the three tokens of place p_1 require $3 \times 3 = 9$ matching cycles for instantiation of (X, Y) and ¬(Y, Z). Naturally the total time consumed for transition tr_1 includes these 9 matching cycles at the LTM. Therefore the total time consumed for firing of tr_1

$$= \text{Time required for 9 matching cycles at LTM + Time required for 1 matching cycle at the Matcher}$$
$$= \text{Time required for } (9 \times 2) = 18 \text{ variable matchings}$$
$$= (18 \times 5) \ \mu S + 1 \times 10 \ \mu S$$
$$= 90 \ \mu S + 10 \ \mu S$$
$$= 100 \ \mu S.$$

Since the preprocessing for firing at tr_2 works in parallel to that of tr_1, tr_2 does not consume any additional time. Thus the time required for firing tr_1 and tr_2 together

$$= 100 \ \mu S.]$$

6. Consider the Local Token Matcher in Fig. 4.6. The doted box in the Fig. 4.6 against which the label is attached, contains two components A and B. A denotes the sign of the arc function variables and the first component of B denotes the sign of the token in the associated place of the transition. The last three components of B contains the values of variables X, Y and Z respectively.

    Let AF_1 and AF_2 denote the registers containing respective sign and value of arc functions (X, Y) and ¬(Y, Z) respectively. Assuming that the time required for sign matching is 2 µS, value matching is 3 µS and propagation delay for 1 AND gate is 1 µS, determine the time needed to obtain a consistent binding of the two arc function variables.

    [**Hints:** The sign and value matching here can be done in parallel in Fig. 4.6. We note that the time needed for value matching is more than that of sign matching, the total detrimental factor in timing is given by the values only. Hence the time needed to generate a consistent binding, if possible

    = Max (time required for sign matching and value matching)
    + (3 AND-delay)
    = Max (3 µS, 2 µS) + 3 × 1 µS
    = 3 µS + 3 µS
    = 6 µS.]

7. Consider the TSF in Fig. 4.10. The set of the current and the used bindings for transition tr_1 of Fig. 3.7 in the second firing cycle is listed below.

    Set of current bindings: {(r/X, n/Y, a/Z)/(r/X, d/Y, a/Z)}
    Set of used bindings: { {r/X, d/Y, a/Z}, {r/X, n/Y, a/Z} }

    The time required for matching of three variables X, Y and Z in parallel = 3 µS, determine (a) the master clock frequency. Also (b) determine the time required for generating the flag from the TSF.

    [**Hints:**

    (a) The master clock should be sufficiently wide to hold the address at the input of the MUX for comparison of its corresponding content with the Set of current bindings (SOCB).

    Given the time required for concurrent matching of three variables, we can determine the clock frequency

    = 1/ (3 µS)

$= 0.33 \times 10^6$ Hz
$= 0.33$ MHz.

(b) For the generation of the flag it is needed to check whether the current binding is a subset of used bindings. In the present context it requires two matching cycles of three variables. Thus the time needed for generation of flag $= (2 \times 3)\ \mu S = 6\ \mu S.]$

References

1. Hwang, K. and Briggs, F. A., *Computer Architecture and Parallel Processing*, McGraw-Hill, Singapore, 1986.
2. Konar, A., *Uncertainty Management in Expert Systems Using Fuzzy Petri Nets*, Ph.D. thesis, Jadavpur University, 1994.
3. Mano, M. M., *Computer System Architecture*, Prentice-Hall, Englewood Cliffs, NJ, 1982.
4. Mathur, A. P., *Microprocessors*, Tata McGraw-Hill, 1985.
5. Russel, S. and Norvig, P., *Artificial Intelligence: A Modern Approach*, Prentice-Hall, Englewood-Cliffs, NJ, 1994.
6. Silberschatz, A. and Galvin, P. B., *Operating System Concepts*, Addison-Wesley, Reading, MA, 1994.

5

Parsing and Task Assignment on to the Proposed Parallel Architecture

The chapter provides an outline to parsing users' pseudo PROLOG codes and mapping segments of the program onto the parallel architecture introduced in the last chapter. To avoid online mapping of the program segments onto the architecture, we presume no constraints on system resources. If the situation is different, then a specially designed task-assignment policy is needed to identify the usable (non-utilized) hardwired resources and a dynamic task assignment program is to be invoked for mapping the non-executed program modules onto the usable system resources. It is indeed important to note that such type of online mapping of system resources causes a significant delay in the execution of the program. A speed-size tradeoff is commonly used to optimize the size of the architecture and minimize the execution time of the program. The architecture we presented in the last chapter is economized based on the resource demand in the execution phase of the program. Consequently, no online mapping of resources is needed in the present context.

5.1 Introduction

This chapter provides details of the pre-processing needed prior to execution of a pseudo PROLOG program. Classical PROLOG programs like programs in traditional languages such as C or Pascal are first compiled to obtain the target object code for running them on a given processor. The processor in turn generates its micro-codes for execution of the object codes from a micro-programmable ROM, and issues appropriate control signals for bus opening and closing for data transfer and execution. A number of alternative execution models of PROLOG programs (vide Fig. 5.1) are prevalent in the current realm of Artificial Intelligence. Most of the models, however, include transformation of a PROLOG program into an intermediate code suitable for *Warren Abstract Machine (WAM)* [5,7], and then translation of the WAM code into the object code of the host machine. The WAM provides a framework for automatic detection of concurrency in a PROLOG program, thereby facilitating the users to run the concurrent object codes on a parallel architecture.

A. Bhattacharya et al.: *Parsing and Task Assignment on to the Proposed Parallel Architecture*, Studies in Computational Intelligence (SCI) **24**, 211–228 (2006)
www.springerlink.com   © Springer-Verlag Berlin Heidelberg 2006

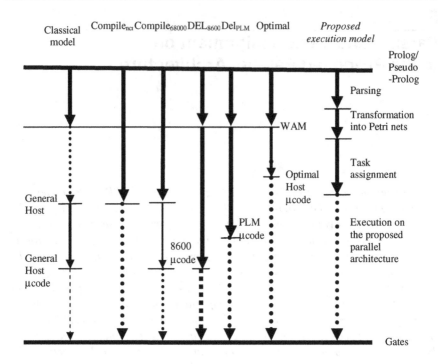

Fig. 5.1: A comparison of various alternative execution models of PROLOG including the proposed model

The chapter proposes a new model for execution of a pseudo PROLOG program (vide rightmost part of Fig. 5.1). Here, the program is first passed on to parser for lexical analysis and syntax checking. An emulation program then transforms the bug-free source code into a Petri net. Hardware resource requirements for execution of the program thus can be traced from the emulated Petri net model. A task assignment program then determines the hardware requirements from the emulation model and maps the program resources onto the hardwired system resources. For instance, the binary coded place names are mapped to the associated place fields (APp_i) of the Transition History File (THF). The signed arc functions detected from the Petri net model of the pseudo PROLOG program are also mapped to the arc function associated with a place ($AFAWp_i$) of the THF. The signed tokens are mapped at the place buffers of the appropriate Place Token Variable Value Mappers (PTVVMs). The current and used binding fields of the Transition Status File (TSF) are assigned null values prior to execution of the program.

One point that needs to be addressed here is whether to add flavor to static or dynamic assignments. In case of static assignments, the program resources can be mapped onto the architecture once only and prior to the execution of the program. Dynamic assignment [4] on the other hand is complex as it involves many issues like deciding usability of system resources and their dynamic interconnectivity. On occasions, there exist options in resource assignments, and consequently an optimal policy in assignment strategy is framed to resolve the issue. Both deterministic and stochastic models [2] are employed in policy design, and the choice of the policy greatly depends on usable resource count, their type and the user resources such as arc function variables, tokens and predicates. To keep our design simplified, we in this chapter do not consider dynamic assignment of program resources onto the proposed architecture.

It also needs mention here that the First Pre-condition Synthesizer (FPS) introduced in the last chapter generates a flag based on the existence of tokens at all the places, excepting at most one. The pre-processor provides the necessary logic function to the FPS to check the above status, which is subsequently synthesized by the FPS for its realization. The flag to indicate the status of the FPS is then generated for satisfying the subsequent firing condition of the transition under reference.

5.2 Parsing and Syntax Analysis

Parsing is a fundamental step in the process of translation/compilation of a source code to a target code. It ensures correctness in the source code and performs lexical and syntactic analysis to detect the bugs in the source codes. The program that is used for parsing is called a **Parser**. Parsers usually require a grammar to check the correctness of the source codes. The source code in the present context is in pseudo PROLOG format of the following type.

Sample Program

Symbols X, Y, Z, a, b, c.
Predicates P (X, Y), Q (Y,Z), R (X, Y, Z).
Clauses

$$P (a, b) \leftarrow. \tag{5.1}$$
$$Q (b, c) \leftarrow. \tag{5.2}$$
$$R (X, Y, Z) \leftarrow P (X, Y), Q (Y, Z). \tag{5.3}$$
$$P (X, Z), Q (X, Y) \leftarrow R (X, Y, Z). \tag{5.4}$$

It is clear from the sample pseudo PROLOG program presented above that the *symbols* including variables X, Y, Z and constants like a, b and c are separated by commas and terminated by a period (.). Further, the *predicates* are also separated by commas and the last predicate is identified from the period following it. The *clauses* are separated by a period, and literals in the body of a clause are separated by comma. Each clause is terminated by a period. The main difference of the present pseudo code with respect to a PROLOG code is that the current coding allows more than one literal in the head.

Quite a large number of parsing techniques is available in the current literature of compilation [1, 3]. Among these *parse tree-* and *deterministic finite automata (DFA)–*

based schemes need special mention. An illustrative tree based scheme for parsing is presented in Fig. 5.2. Given a pseudo PROLOG statement, a parse tree is constructed by gradually expanding a clause using the re-write rules. Usually the re-write rules are constructed judiciously to describe all possible vocabulary of a given language, and they together are called the grammar of the language. A simple grammar to test the syntax of a pseudo PROLOG language is presented as follows:

Grammar

$$\text{Symbols} \Rightarrow X \mid Y \mid Z \mid a \mid b \mid c \tag{5.5}$$
$$\text{Predicates} \Rightarrow (\text{Pre (arg1, arg2)}) \, (,\text{Pre (arg1, arg2)})^{*} \tag{5.6}$$
$$\text{arg1} \Rightarrow \text{Symbols} \tag{5.7}$$
$$\text{arg2} \Rightarrow \text{Symbols} \tag{5.8}$$
$$\text{Pre} \Rightarrow P \mid Q \mid R \tag{5.9}$$
$$\text{Head} \Rightarrow \epsilon \mid \text{Pre (arg1, arg2)} \mid \text{Pre} \tag{5.10}$$
$$\text{Tail} \Rightarrow \text{Pre} \mid \text{Pre (arg1, arg2)} \mid \text{Predicates} \tag{5.11}$$
$$\text{Clause} \Rightarrow \text{Head} \leftarrow \text{Tail Period} \mid \text{Head} \leftarrow \text{Period} \tag{5.12}$$
$$\text{Query} \Rightarrow \leftarrow \text{Tail Period} \tag{5.13}$$

where the upper case letters like X, Y and Z stand for variables, and lower case letters like a, b, c represent constants;

"|" denotes an OR operator;

"⇒" sign stands for a replacement operator. This in other words means that the left hand side of the re-write rule can be replaced by its right hand side;

"*" above a symbol here denotes one or more number of occurrences of the symbol, and

"ε" denotes a null string.

5.2.1 Parsing a Logic Program using Trees

This section provides a discussion on the construction of parse trees for statements in a pseudo-PROLOG program. As already discussed, the parse tree is constructed by expanding a clause using the grammar supplied. The left-hand side of the re-write rules is compared with the available string or its part located at a node (initially at the root) of tree. A rule whose left hand side matches with the given clause is selected, and the matched clause in the source code is replaced by the right-hand side of the selected rule. The updated clause is placed in the tree as an offspring of its previous form. The nodes in the parse tree are thus expanded until the whole source code appears at the leaves in the left-to-right traversal of the tree.

Both top-down[1] and bottom-up parsing can be employed for the construction of parse trees. In this section, we illustrate top-down parsing (vide Fig. 5.2) with respect to the following pseudo PROLOG statement.

$$\text{Grandfather } (X, Z) \leftarrow \text{Father } (Y, Z), \text{Father } (X, Y). \qquad (5.14)$$

In fact any typical top-down parser such as L-R parser [1] that offers predictive matching capability and is free from back-tracking is suitable for our purpose. Since the algorithms for such parsers are available in any standard text on compiler [1, 3], we for the sake of brevity omit its discussion here.

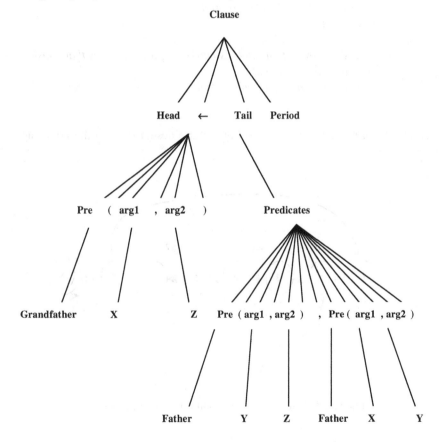

Fig. 5.2: Parse tree used to analyze the syntax of a user defined clause:
Grandfather (X, Z) ←Father (Y, Z), Father (X, Y)

---

[1] In top-down (bottom-up) parsing, the tree is expanded from the root (leaves) and construction of the tree is continued until the whole expression is completely parsed.

5.2.2 Parsing using Deterministic Finite Automata

Instead of a parse tree, a DFA can equally be used to handle the parsing problem of the pseudo PROLOG codes. A *DFA* is defined as a 5-tuple, denoted by

$$DFA = \{S, N, A, T, E\}$$

where

        S is the start symbol denoted by an arrow to a state,
        N is the set of states denoted by circles,
        A is the set of arcs,

        T is the set of transition symbols marked against the arcs that cause a
          transition of states, and
        E is the termination/ end symbol, denoted by 2 concentric circles.

The DFAs presented in Fig. 5.3 to Fig. 5.9 are designed using the grammar defined earlier.

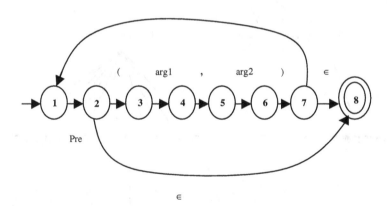

Fig. 5.3: A finite automation for tail part of a given clause

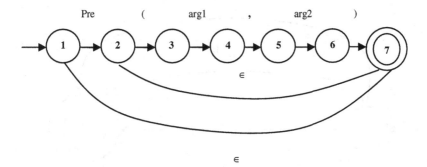

Fig. 5.4: A DFA for the head part of a clause

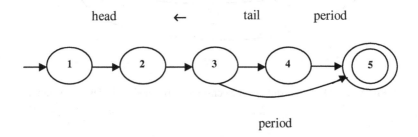

Fig. 5.5: A DFA of a goal / general clause

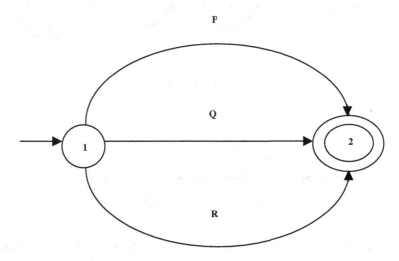

Fig. 5.6: Finite automation of Pre following definition (5.9)

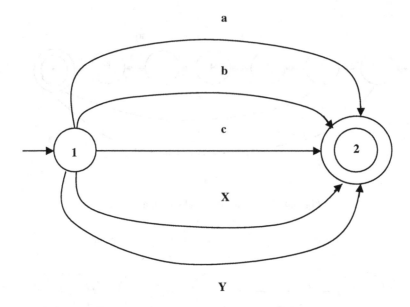

Fig. 5.7: A DFA for symbols like constants a, b, c and variables like X and Y

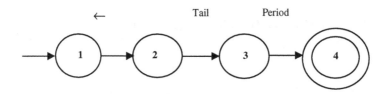

Fig. 5.8: A DFA for a query

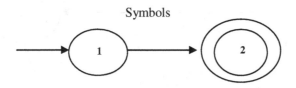

Fig. 5.9: The Argument-DFA for the rules: arg1 $\Rightarrow$ Symbols and arg2 $\Rightarrow$ Symbols

To illustrate the implication of the above automata in parsing a given statement, let us consider the following pseudo PROLOG query:

Query: $\leftarrow$P (X, Y), Q (Y, Z). (5.15)

Now when the query (5.15) is submitted in runtime, the control automatically consults the automation of a standard query (Fig. 5.8). The implication sign ($\leftarrow$) is matched with the arc connected between state 1 and state 2 of Fig. 5.8. If a successful match occurs, then a DFA for tail matching is explored. So, a switching, demonstrated by a dotted line, occurs from state 2 of a query-DFA to state 1 of the tail-DFA. Again, when a Pre is encountered in the tail-DFA, a transition takes place following state 1 of tail-DFA to state 1 of Pre-DFA, and on matching of the appropriate predicate symbol in the Pre-DFA, a further transition to state 2 of tail-DFA takes place. The left parenthesis "(" of the input string automatically matches in the arc between state 2 and 3 of the tail-DFA. Then on encountering an argument in the input string, a transition following state 3 of tail-DFA to the argument-DFA (for lack of space in Fig. 5.10) takes place. The argument-DFA calls the symbol-DFA for term (variables/ constants) matching of the input string with those enlisted in the symbol-DFA. The control then returns from the symbol-DFA to Argument–DFA (not shown) and from the argument-DFA to state 4 of the tail-DFA. The rest of the matching in the tail part of the input string is obvious. If the tail matching is alright, a transition from the end state of the tail-DFA to state 3 of query-DFA takes place. On finding a successful match of a period of the input query with the same in the query-DFA, the parser accepts the query for subsequent evaluation. The method of testing any other clause is analogous and thus is not discussed in detail.

5.3 Resource Labeling and Mapping

Prior to initiate reasoning on the architecture proposed in chapter 4, the resources of the architecture should be properly labeled. The resources in the present context are of two distinct types: (i) hardwired resources, and (ii) simulated Petri net resources such as arc function variables and constants. The hardwired resources include PTVVM, TSF, MATCHER, FPS and THF and Firing Criteria Testing Logic (FCTL). The modules of the simulated Petri net are mapped onto appropriate hardwired resources. For convenience in subsequent operations, hardwired and arc function resources are named in binary codes before the mapping is accomplished.

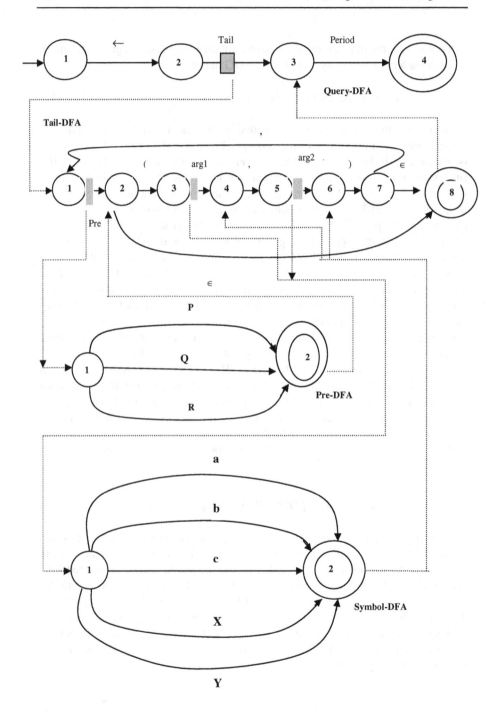

Fig.5.10: A complete DFA for testing the syntax of a query statement

5.3.1 Labeling of System Resources

Once parsing is over and the logic program is detected to be free from syntactical errors, the pre-processing program starts labeling the system resources. In order to label the places and transitions, we need to construct a pseudo Petri net from the given program clauses. The number of places found in the Petri net is first counted, and consequently the places are named by binary numbers. In case there are n places in the pseudo Petri net, the places are numbered from 0 to (n −1) in binary code, so that all places have a distinct place name. The number of transitions in the Pseudo Petri net are also counted and numbered similarly.

For identifying the variables in an arc function, the arguments of each predicate are identified separately in the process of parsing. A buffer is then initialized with each transition, and the variables associated with a transition are saved in the buffer in a manner so that the buffer does not contain the multiplicity (multiple copies) of a variable. The variables thus saved in the buffer are counted and named in binary code so that all of them have a distinct binary variable name. Suppose there are altogether m buffers. Once the naming of variables of the first buffer is over, the variable naming of the second buffer starts. Thus if the last variable taken from buffer 1 is given a variable name r (in binary code), then the first element of the second buffer will have a variable name r + 1. The rest of the variables of the buffer 2 will thus have distinct variable names r + 2, r + 3, etc. The process of variable naming is thus continued until the last variable of the last buffer is exhausted.

5.3.2 The Petri Net Model Construction

Petri net model construction is an intermediate step in mapping program clauses on to the hardwired resources. In fact this intermediate step has a number of advantages, specially in planning and organizing the hardwired resources according to the user's need. The construction of Petri net is accomplished first by transforming *head ←tail period* type program clauses on to a sub-net. The algorithm used for the above mapping is trivial. It checks, whether there exists any transition in the net with input places corresponding to the enlisted predicates in the tail, and output places describing the literal present in the head. If no such transition is found, a new is created. Places labeled with a predicate name same as in the head or tail are then searched in the existing sub-net so far created. If one or more predicates enlisted in the clause are absent, new places are created and the places are labeled with appropriate predicate names following the current list of unfound literals associated with the given program clause. When all literals of the given clauses are available as places, the places are attached to the transition as input or output places depending on their presence in the body or head parts of the clause respectively. Arguments of each predicate are then labeled as arc functions against the arcs connected between the appropriate place containing the same label and the transition under reference.

The process of mapping program clauses on to the Petri net is continued until the whole set of clauses, described above, are transformed into Petri net construct. The atomic program clauses, having only a head literal, are now searched in the Petri net, and the arguments of such clause are mapped as token at the appropriate place containing the same predicate label. The sign of arc functions are also labeled, depending on their association with the arc types such as place to transition arcs or transition to place arcs of the selected clause. For the former type the sign should be positive, and for the latter case it is negative.

On completion of the Petri net construction, the hardwired resources can directly be mapped from the simulation model of the Petri net, rather than mapping the same from the user-supplied source code.

5.3.3 Mapping of System Resources

The register files in the architecture of the proposed scheme needs to be initialized before the hardwired execution of the reasoning program starts. For instance, the place names are initialized at the appropriate register file associated with a transition. In other words place names are mapped to the APp_i fields of the transition history register files.

The signed arc functions are first detected from the program clauses in the process of parsing, and saved in temporary buffers. Later these signed arc functions are transferred from the temporary buffer to the appropriate $AFAWp_i$ fields of the transition history file.

For determining the signed constants in the given clauses, the parser should identify the body-less clauses (i.e. clauses with a head, an arrowhead and a period only) and the sign of the tokens, which is explicitly available within the program clause. For example, the signed token part in the body-less program clause: $\neg$Likes $(r, l) \leftarrow$. is $\neg$<r, l>. These signed tokens are mapped at the place buffers of the PTVVM.

The current and used instantiation fields of the TSF are also initialized with null values prior to initiate the execution of the program.

The compiler also synthesizes the logic for testing one of the firing conditions of a transition. The condition in the present context refers to checking whether all but one place associated with a transition possess tokens. As an example, let a transition tr_i has r number of input places $p_1, p_2, ..., p_r$ and s number of output places $p_{r+1}, p_{r+2}, ..., p_{r+s}$. Then the condition that ensures tokens at all but one place is presented below.

$$(p_2 \; p_3 \ldots .p_r) \; (p_{r+1} \; p_{r+2} \ldots .p_{r+s}) + (p_1 \; p_3 \ldots .p_r) \; (p_{r+1} \; p_{r+2} \ldots .p_{r+s}) + (p_1 p_2 \; p_4 \ldots .p_r)$$
$$(p_{r+1} \; p_{r+2} \ldots .p_{r+s}) + \ldots + (p_1 p_2 p_3 \ldots .p_{r-1}) \; (p_{r+1} \; p_{r+2} \ldots .p_{r+s}) + (p_1 p_2 \; p_3 \ldots .p_r)$$
$$(p_{r+2} \ldots .p_{r+s}) + (p_1 p_2 p_3 \ldots .p_r) \; (p_{r+1} p_{r+3} \ldots .p_{r+s}) + \ldots + (p_1 p_2 \; p_3 \ldots .p_r) \; (p_{r+1}$$
$$p_{r+2} \ldots .p_{r+s-1}) = \text{True} \qquad\qquad (5.16)$$

In the above Boolean condition, $p_i = 1$ denotes that the place p_i possesses tokens, and obviously $p_i = 0$ indicates that place p_i has no tokens. *Sum* and *product* operators in the last condition represent Boolean OR and AND operations respectively.

After the compiler generates the above condition, it is passed on to the First Pre-condition Synthesizer (FPS) logic that implements the above logic for verifying the pre-condition for transition firing. In fact thope FPS includes a set of logic gates that is automatically configured to satisfy the desired logic function.

5.4 Conclusions

Prior to execution of a pseudo PROLOG program on the proposed architecture, the source code needs to be parsed and the symbols extracted from the source code are required to be mapped onto the said architecture. In the present context, a *deterministic finite automation* was employed to check the syntactical errors in the source code during the process of parsing. In case the source code is free from syntactical error, the pertinent parameters of the program such as arc function variables are represented by distinct binary numbers and mapped at the appropriate units in the architecture. In fact a pseudo Petri net is created by the compiler to trace the places and transitions in the architecture with respect to those in the Petri net. Such correspondence helps in identifying the fired transition and the resulting bindings easily for answering a user's query. In our elementary design, we do not consider construction of a symbol table to hold the program variables, as we have a limited number of variables in example programs. However, for practical systems, symbol tables [6] need to be constructed to determine the location of the symbols in the memory. Symbol table construction is not discussed here as we worked with limited number of variables.

Exercises

1. Construct the grammar and build a parse tree using the grammar for the following logic program:

    Variable: X, Y;
    Constant: a, b, c;
    Predicates: P(,), Q(,);
    Clauses : P(X, Y) ← Q(Y, X).;
            Q(a, b) ←.;
            Q(b, c) ←.;

    Query   : ← P(a, c).;

[Hints:

Grammar:

Symbols ⇒ X│Y│a│b│c
Predicates ⇒ (Pre(arg 1, arg 2))(, Pre(arg 1, arg 2))*
arg 1 ⇒ Symbols
arg 2 ⇒ Symbols
Pre ⇒ P│Q
Head ⇒ ∈ │Pre(arg 1, arg 2)│Pre
Tail ⇒ ∈ │Pre(arg 1, arg 2)│Pre│ Predicates
Clause ⇒ Head ←Tail Period│Head ←Period
Query ⇒ ← Tail Period

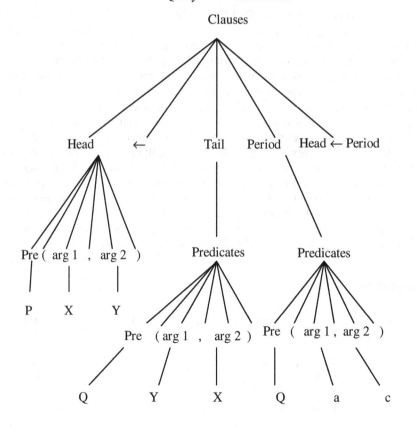

Fig. 5.11: A parse tree

A parse tree is constructed using the grammar vide Fig. 5.11   ]

2. Using the deterministic finite automata for clause, head/tail part of clauses, predicates, variables/constants, queries and symbols, as given in the text represent the following clause by deterministic finite automata.

$$P(X, Y) \leftarrow Q(Y, X).$$

[**Hints:** A deterministic finite automata is constructed to represent the given clause vide Fig. 5.12.

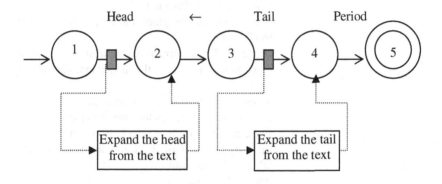

Fig. 5.12: A DFA for a clause]

3. (a) Given a set of clauses and facts, construct an algorithm for representing them using a Petri net.

   (b) Defining the necessary parameters of a given program, evaluate the complexity of your algorithm.

[**Hints:**

(a) The algorithm for Petri net construction greatly depends on the list of rules and facts. The algorithm begins with scanning literals and operators in the rule one by one, until the rule is terminated by a period or semicolon. Since the rules are scanned following the grammar, the types of the literals such as predicates, arguments etc. can easily be identified. After detection of each predicate, the same is searched as a place name in the Petri net so far constructed. If the place of the desired name is not available in the Petri net, then the Petri net is augmented with the new place name. When all the predicates of the rule are found to be available on the Petri net, a transition is constructed to connect the predicates in the antecedent part to the predicates in the consequent part by placing one transition in between. A simple algorithm that serves the above need is presented below.

Procedure EPN-construction
Begin
  While no-of –rules is not exhausted **do**
    Begin
    For each rule
      If the predicates found in the rule is not available in the Petri net
      Then construct the place and label it to augment the Petri net
      If the predicate corresponding to the place now constructed belongs to the antecedent part of the rule
      Then mark the place as the input place
      Else mark the place as the output place
      Place a transition to connect the input places to the output places of the transition and attach a label to the transition
      Attach Arc-function
    End-for;
  End-while;
End.

Now to attach the arc functions with the arc and to map the tokens in the places, we need to consider the facts available in the system. The following procedure may then be invoked to attach arc functions and map tokens in the places.

Function Attach Arc-function

Begin
  Identify the tokens associated with the predicates and according to the type of the corresponding places attach a suitable sign (+ for arc function corresponding to the input places and – for the arc function corresponding to the output places);
  Also attach the arguments of the predicates in order with the arc containing the place and the transition under reference.
End.

Function Attach Tokens

For each fact
Repeat
 Isolate the argument of the predicate representing the fact and map it with the
 given sign at the corresponding place in the EPN.
End.

(b) Let a_i be the no. of predicates in the antecedent part and c_i be the no. of
predicates in the consequent part of the ith rule. Let the maximum no. of
places in the EPN be N. Now to map a place into the EPN, we atmost
need N comparisons. Thus to map $a_i + c_i$ no. of places, we require $(a_i + c_i)$
N number of searches in the worst case. Let the number of rules be M.
Thus for mapping M rules on to the EPN we need at most

$$\sum_{i=1}^{M} (a_i + c_i) \text{ N comparisons.}$$

Let $a_i \leq A$ and $c_i \leq C$ for all i =1 to M.
Thus the above result reduces to

$$MN (A + C).$$

Thus the complexity is

$$O (MN).$$

To identify an arc for mapping an arc function we need MN searches.
Thus for all clauses together the total search cost will be MN $(A + C)$.

For mapping the tokens we need to identify the place on the EPN. Since
there N places, the total search cost for mapping $(A + C)$ no. of tokens we
need $N(A + C)$ search cost. Thus for M rules the search cost becomes MN
$(A+ C)$.

Summing up the three search costs we find the order of complexity
remains MN $(A + C) \approx O(MN)$ since $(A + C)$ is much negligible in
comparison with MN.]

4. For the following logic program,

(a) Encode the system resources into appropriate binary strings, and outline the mapping of these strings on to appropriate modules of the architecture.

(b) Which information during the execution of the program are mapped online onto the architecture?

Logic Program:

Son(Y, X) ←Father(X, Y).
Father(d, r) ←.

[**Hints:**

(a) The number of places and transitions required to realize the given logic program are first determined. The necessary word-length of the appropriate units is then fixed in the respective architecture. The variables, arc functions and their signs are encoded into binary strings of appropriate length. These are then mapped onto appropriate modules of the architecture for initialization of the execution process.

(b) The current variable bindings and the set of used bindings which are computed online are also mapped onto appropriate registers in real time.]

References

1. Aho, A. V., Sethi, R., and Ullman, J. D., *Compilers: Principles, Techniques and Tools*, Addison-Wesley Longman, Singapore, 1999.
2. Culler, D. E. and Singh, J. P., *Parallel Computer Architecture: A Hardware/Software Approach*, Morgan-Kaufmann, CA, 1999.
3. Dhamdhere, D. M., *Systems Programming and Operating Systems*, Tata McGraw-Hill, New Delhi, 2002.
4. Hwang, K. and Briggs, F. A., *Computer Architecture and Parallel Processing*, McGraw-Hill, Singapore, 1986.
5. Patt, Y. N., "Alternative Implementations of Prolog: the micro architecture perspectives," *IEEE Trans. on Systems, Man and Cybernetics*. vol. 19, no. 4, July/August 1989.
6. Tremblay, J-P. and Sorenson, P. G., *An Introduction to Data structures with Applications*, McGraw-Hill, NY, 1984.
7. Yan, J. C., "Towards parallel knowledge processing," in *Advanced series on Artificial Intelligence, Knowledge Engineering Shells: Systems and Techniques*, Bourbakis (Ed.), vol. 2, World Scientific, Singapore, 1993.

6

Logic Programming in Database Applications

The chapter addresses the scope of logic programs in database systems. It begins with syntax and semantics of the Datalog language, and highlights the special features of the language in answering database queries. The LDL system architecture, which until this date is the most popular system for execution of Datalog programs, is then briefly introduced. Next the scope of Petri net models in designing database systems is examined. The techniques to overcome the limitations of the LDL system architecture by Petri net models are also presented. The chapter ends with a discussion on the use of Petri net based models in Data Mining applications.

6.1 Introduction

The book introduces the scope of parallel and distributed logic programming with a special emphasis on the architectural issues of logic program machines. This chapter deals with Datalog language, which has proved its significance in database applications. The chapter attempts to justify the significance of the proposed logic programming machine for realization of Datalog programs.

The chapter begins with an introduction to Datalog language. The semantics of Datalog program and the principles of answering user-made queries in Datalog language have been outlined and illustrated with many examples. The representational benefit of integrity constraints in Datalog programs is also illustrated. The LDL system architecture is taken as a case study to understand the execution of a Datalog program on a practical system. The scope of Petri net models in designing database machines is also outlined in this chapter. Finally the chapter stresses the need of data mining on the modified architecture supported by Petri nets.

6.2 The Datalog Language

Datalog [1-25] is one typical logic program based query language that supports the formalisms of Horn clause based logic programs. The syntax of Datalog resembles

A. Bhattacharya et al.: *Logic Programming in Database Applications*, Studies in Computational Intelligence (SCI) **24**, 229–258 (2006)
www.springerlink.com

the syntax of Prolog. This section presents the basic structure of Datalog programs and the principles by which the answer to a query is determined.

Consider for instance an account relation consisting of three attributes branch-name, account-number and balance, as shown in table 6.1.

Suppose, we design a view relation: view1 that contains account-number and balances for the accounts at the Park Circus branch with a balance more than Rs 7000. The following rule is a representative description of the given problem:

$$\text{view1}(A, B) \leftarrow \text{account}(\text{Park Circus}, A, B), B > 7000. \tag{6.1}$$

To retrieve the balance of account-number 0201 we use the following query:

$$\leftarrow \text{view1}(0201, B). \tag{6.2}$$

The answer to the above query can automatically be generated by resolution of the clauses (6.1) and (6.2), which yields:

$$\leftarrow \text{account}(\text{Park Circus}, 0201, 8500).$$

Table 6.1: The 'account' relation.

Branch-name (N)	Account-number (A)	Balance in rupees (B)
Jadavpur	0101	5000
Sodpur	0215	1000
Maniktala	0102	9000
Park Circus	0322	6500
C R Avenue	0305	4500
Park Circus	0201	8500
Park Street	0222	7500

Thus account-number 0201 in Park Circus branch has a balance of Rs 8500.

Now suppose we want to identify the account numbers whose balance is > 7500. This can be represented by the following query:

←view1(A, B), B > 7500.

The owner to the above query can directly be obtained from the account relation. The answers to the query are given below:

←account(Maniktala, 0102, 9000).
←account(Park Circus, 0201, 8500).

The Datalog programs are also capable to compute answers of a query not directly available in the given relation (table 6.1). For example, let us consider the problem of computing interest of an account using the following rule:

Rule1: If balance < 2000, then interest-rate = 0%.

Rule2: If balance ≥ 2000, then interest-rate = 4%.

These types of rules can be coded in Datalog program as outlined below:

interest-rate(A, 0) ←account(N, A, B), B < 2000.
interest-rate(A, 4) ←account(N, A, B), B ≥ 2000.

Using the above rule, we can evaluate interest of all the account numbers cited in the account relation.

Sometimes negation is also used in a Datalog program [20]. For example, let us try to construct a Datalog program to identify all customers in a bank who have a deposit but have no loans. The following Datalog program serves the purpose:

customer(N) ←depositor(N, A), **not** is-borrower(N).
is-borrower(N) ←borrower(N, L).

where N, A, L denote customer-name, customer account-number and customer loan-number respectively.

We have presumed that the relations depositor(N, A) and borrower(N, L) are available in table 6.2 and 6.3 respectively.

Table 6.2: The 'depositor' relation.

Name (N)	Loan (L)
ram	0101
shyam	0200
jadu	0021
sita	0012
mira	0002
kali	0202

Table 6.3: The 'borrower' relation.

Name (N)	Account-number (A)
ram	0101
shyam	0200
jadu	0021
madhu	0201
ganga	0120
hari	0221

The answer to the given query:

$$\leftarrow customer(N).$$

with respect to the above relation are presented below:

    customer(madhu),
    customer(ganga),
    customer(hari).

6.3 Some Important Features of Datalog Language

From a syntactic point of view the positive and negative literals are represented in the following format

$$p(t_1, t_2, \ldots\ldots, t_n)$$

$$\text{not } p(t_1, t_2, \ldots\ldots, t_n)$$

where the relation p has n attributes: $t_1, t_2, \ldots\ldots, t_n$.

Algebraic operations such as summation or subtraction can be represented in the Datalog language as relations. For example,

$$X = Y + Z \qquad \text{can be represented in Datalog as}$$

$$+ (Y, Z, X).$$

where the '+' denotes a relation of three attributes X, Y and Z.

Datalog language supports Boolean conditions using $>$, $=$, $<$ relations. One example program indicating the use of algebraic and Boolean relation is presented below to illustrate the computation of banking-interest in a given branch of a bank.

Example 6.1: Consider the relations given in connection with accounts and interest for each account in a given bank.

Relations

interest(A, I) represents the interest I for the account-number A.
account-status(A, B) indicates the balance B for the given account-number A.
interest-rate(A, R) denotes the rate of interest R over an account-number A.
account(N, A, B) denotes an account of person N having account-number A with a balance B.

The logic program below is developed to determine the interest of all the accounts in the "Jadavpur" branch.

interest(A, I) ←account-status(A, B), interest-rate(A, R), I = B * R/100.
account-status(A, B) ←account("Jadavpur", A, B).
interest-rate(A, 0) ←account(N, A, B), B < 5000.
interest-rate(A, 4) ←account(N, A, B), B >= 5000.

Another interesting feature of Datalog program is recursive use of relations in the same clause. Example 6.2 illustrates the aforementioned principles.

Example 6.2: Consider the following relations describing employers and managers.

Relations

emp(X, Y) denotes X is an employer of Y.
manager(X, Y) denotes X is a manager of Y.

The following two rules together represent the linkage between employer and manager relations.

emp(X, Y) ←manager(X, Y).
emp(X, Y) ←manager(X, Z), emp(Z, Y).

Here, the employer relation appears in both the left and right hand side of the "←" in the second clause, and hence we say that employer is a recursive relation in the given rule.

One important use of recursion in a Datalog program is the generation of numbers or sequences. For example, the following program generates all even numbers counting from zero.

even-number(2) ←.
even-number(A) ←even-number(B), A = B + 2.

Given a query

'←even-number(N).',

the above program generates the sequence 2, 4, 6, 8, ..., ∞. i.e., the whole set of even numbers.

6.4 Representational Benefit of Integrity Constraints in Datalog Programs

Integrity constraints are usually introduced in a database system to guard or protect accidental damage to the database [22-23]. Representation of integrity constraints in a relational database is not as easy as in logic program based database. The reason behind it lies in the representational advantage of rules/constraints by logic programming languages. Consider for instance, a supplier database consisting of the following relations

supplier(S-no, Name, City-address) indicating the supplier's number, his name and the city-address.

spj(S-no, P-no, J-no, Q) indicating a relation of supplier, part, job and the quantity of part supplied.

job(J-no, J-name, J-city) denotes that a job having number J-no and name J-name has a demand in city J-city.

local-supplier(S) denotes the name of local-supplier, S.

Suppose we want to construct integrity constraints in connection with a supplier-database by suitable Datalog statements.

Statement 1: No local-supplier supplies part p_3.

> **not** spj(S, p_3, _ , _) ←local-supplier(S).

Statement 2: Supplier s_3 supplies every job in Calcutta.

> spj(s_3, _ , j, _) ←job(j, _ , Calcutta).

Statement 3: Supplier s_4 supplies job in Calcutta only.

> job(j, _ , Calcutta) ←spj(s_4, _ , j, _).

6.5 The LDL System Architecture

The current expert database systems are usually realized on loosely coupled architecture [21]. Such systems consist of a front end and a back end, where the front end includes domain-specific knowledge and the back end contains general-purpose DataBase Management Systems (DBMS). Users submit queries to the front end. The queries in direct or indirect form then look for the appropriate data on the database through the back end. The results of the queries are transferred back to the users through the front end. A schematic view of a typical loosely coupled architecture is given in Fig. 6.1.

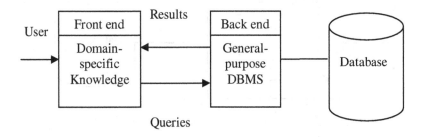

Fig. 6.1: A loosely coupled expert Database system

The loosely coupled architecture introduced earlier suffers from a number of limitations:

(1) There always exists a semantic mismatch between the front-end programming language and the back-end database systems. Usually the front-end language embodies a procedural programming paradigm in which the solution to a problem is expressed as a sequence of operations on a global state. The back-end database system, in contrast, embodies a declarative programming paradigm in which the solution is expressed in a fashion that describes the problem without specifying the intermediate steps to obtain the results.

(2) The level of granularity of the data objects in the front end and the back end may also have a mismatch in Fig. 6.1. The front-end usually specifies a computation on tuples of data values, while the back end does computation on a set of tuples.

(3) In the realization of the overall system, the implementer is bounded by the data model of the back-end database system. Thus, the front end must be 'tailored' to complement the limitation of the back end.

To overcome the aforementioned limitations, the Microelectronics and Computer Technology Corporation (MCC) provided a new approach to design a tightly coupled architecture for logic programming database system. A schematic view of a tightly coupled system is given in Fig. 6.2.

The system described in Fig. 6.2 does not have clearly defined front-end and back-end components. The mismatch in the object granularity of loosely coupled system is thus removed from Fig. 6.2. Here, a single programming paradigm instead of two paradigms, as in loosely coupled system, is used for query generation and obtaining results from the database. Logic Data Language (LDL),

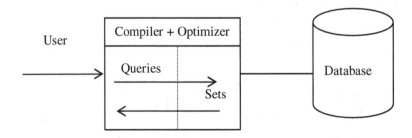

Fig. 6.2: A tightly coupled system

proposed by MCC is one typical language, capable of performing both query generation and management and extraction of results from the database. Some typical characteristics of the LDL are given in order:

(1) The LDL is a declarative language that employs logic programming for data manipulations and management.

(2) The LDL has a fixed point semantics based on the notion of bottom-up query computation.

(3) The data-model in LDL is quite rich as it includes atoms, complex objects, lists and sets of objects.

(4) The LDL is enriched by full negation support and constraint specification capability in the form of equality and inequality predicates.

(5) The LDL also supports compilation techniques for semantic analysis and optimization of user-supplied queries.

(6) A procedural capability for updating is an inherent feature of the LDL system.

(7) The LDL system provides a convenient interface for higher application specific interfaces.

In the next section we examine some of the interesting features of the LDL system.

6.5.1 Declarative Feature of the LDL

As already introduced earlier, the LDL system provides a declarative linguistic support, i.e., user need not specify the detail steps in solving a problem, but needs to mention the problem only using relational constraints [8]. Example 6.3 illustrates the declarative feature of the LDL.

Example 6.3: Consider, for instance, an employee-relation given by

employee(Name, SSN, Age, Department)

where SSN denotes Social-Security-number.

Suppose that the user generates a query: Determine all employee-names working in the hardware department having age below 30 years.

The above query can be represented in a declarative manner as

?{employee.Name | employee.Department = hardware, employee.Age < 30}

The procedure to evaluate the query is left to the system and hence the LDL system is said to be declarative.

It is indeed important to note that the basic Horn clause-based programming style of PROLOG has been extended in the LDL system from the given viewpoint.

A Prolog program just mentions the rules and the facts but leaves the control of program execution with the programmer. In LDL, the control program is automatically invoked to identify the order of selection of the program clauses.

6.5.2 Bottom-up Query Evaluation in the LDL

The answer to a query in a PROLOG program is generated in a top-down left to right order. On the contrary, the query evaluation in the LDL system is accomplished in a bottom-up fashion, starting from the stored database through the relevant rule-bodies to the rule-heads until no new results are produced in the head corresponding to the queries. This form of computation can be formally described as the *fixpoint operator* [21]. The semantics of LDL is described in terms of such fixpoints. Example 6.4 briefly illustrates the comparative merits of the bottom-up evaluation over top-down evaluation.

Example 6.4: Consider the following program clauses describing the definitions of ancestor using the definition of parent.

Cl_1: ancestor(X, Y) ←parent(X, Y).
Cl_2: ancestor(X, Y) ←parent(X, Z), ancestor(Z, Y).
Cl_3: parent(jk, jl) ←.
Cl_4: parent(jk, je) ←.
Cl_5: parent(jl, pr) ←.
Cl_6: parent(jl, my) ←.
Cl_7: parent(my, jn) ←.

The Prolog equation of the above program for the query

?ancestor(jk, X).

can be described as a top-down evaluation following the formalisms of SLD-tree. The SLD-tree shown in Fig. 6.3 presents all the solutions to the problem at the leaves of the tree. On occasions, the tree may be repeatedly expanded by invoking the same rule. To overcome the problem of infinite expansion, the bottom-up approach is adopted in LDL.

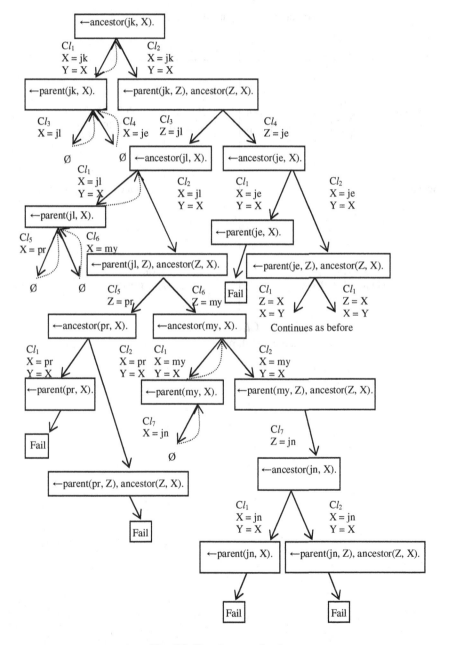

Fig. 6.3: Top-down evaluation

In the bottom-up approach, (Fig 6.4 (a) and (b)) we satisfy the rules by instantiating with the ground literals. The process is repeated until no new solutions are produced.

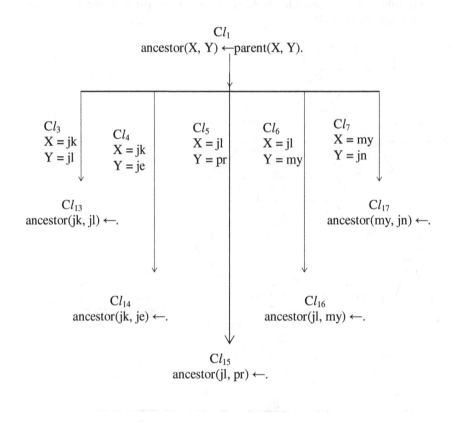

Cl_1
ancestor(X, Y) ←parent(X, Y).

Cl_3
X = jk
Y = jl

Cl_4
X = jk
Y = je

Cl_5
X = jl
Y = pr

Cl_6
X = jl
Y = my

Cl_7
X = my
Y = jn

Cl_{13}
ancestor(jk, jl) ←.

Cl_{17}
ancestor(my, jn) ←.

Cl_{14}
ancestor(jk, je) ←.

Cl_{16}
ancestor(jl, my) ←.

Cl_{15}
ancestor(jl, pr) ←.

(a)

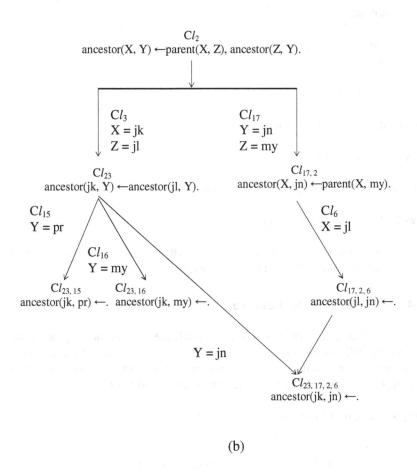

(b)

Fig. 6.4: Bottom-up approach

6.5.3 Negation by Failure Computational Feature

The LDL also supports the well known *Closed World Assumption* (CWA) [13], where literals not supplied as facts are presumed to be false to answer a query. Example 6.5 provides an insight to this problem.

Example 6.5: Given below a logic program, where person (b) is not clearly defined. We thus following CWA presume that ¬person (b) to be true.

Logic Program:

> mammal (X) ←person (X).
> mammal (b) ←.
>
> person (a) ←.

Now suppose user makes the following queries:

Query 1: ? ¬ person(b).

The answer to the query definitely is true.

Query 2: ? mammal (X), ¬ person (X).

The answer to the query definitely is X = b, which is obtained from the direct specification of mammal (b) and absence of person (b).

6.5.4 The Stratification Feature

The stratification feature ensures that every predicate used in the program in its negated form will first be computed in its positive form. Consider, for example, a logic program with stratification over reachability [21] between two nodes in a given graph. Example 6.6 illustrates the stratification used in LDL.

Example 6.6: Consider the following logic program:

> reachable (X, Y) ←edge (X, Y).
> reachable (X, Y) ←edge (X, Z), reachable (Z, Y).
> exclusive-pairs (X, Y, Z) ←reachable (X, Y), ¬ reachable (Z, Y).

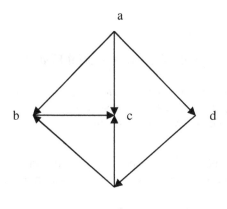

e

Fig. 6.5: A sample graph

where edge (X, Y) denotes that there is a directed edge from node X to node Y, reachable (X, Y) means Y is reachable from node X along some path, and exclusive-pairs derives all pairs of nodes (X, Y) such that Y can be reached from X except for all those pairs that can be reached from node Z.

Figure 6.5 describes a sample graph over which the aforementioned logic program is applied. Suppose we are interested to compute exclusive-pairs (X, Y, d). To answer the query, we first determine edge and reachable relations. By inspection on the graph, the edge relation is found to be

$$\text{edge} = \{(a, b), (a, c), (a, d), (b, c), (d, e), (e, b), (e, c)\}$$
and   reachable = edge $\cup$ {(a, e), (d, b), (d, c)}.

It is important to note that exclusive-pairs include all reachable relations except {(d, e), (d, b)}. Thus,

$$\text{exclusive-pairs} = \text{reachable} - \{(d, e), (d, b)\}.$$

6.6 Designing Database Machine Architectures using Petri Net Models

The LDL language provides a historical landmark on logic program based data language [26-27] for efficient execution of database programs on specialized database machines. Undoubtedly LDL on many circumstances could outperform the traditional PROLOG based logic program machines. The Petri net approach to logic programming introduced in the text, however, provides a more elegant architecture for high performance database machines. Here, unlike Horn clauses, we can directly execute predicate logic based syntax with multiple literals in the head of a clause. The top-down or bottom-up approach for execution of a program is irrelevant in the proposed architecture. The most interesting feature of the Petri net based architecture is that the program clauses need not wait for resolution, rather when suitable data clauses are available concurrent resolution takes place among all the clauses associated with a transition. This particular feature of Petri net based machine significantly enhances the computational speed for the next generation database machines. The negation by failure and stratification feature can easily be implemented in the proposed Petri net based database machines. To illustrate the computational power of a database program on a Petri net based architecture, let us consider, example 6.7.

Example 6.7: The ancestor finding problem introduced in example 6.4 is solved here using the Petri net approach (vide Fig. 6.6).

Here, following Procedure Automated-Reasoning of chapter 3, we first fire transition tr$_1$ generating the following tokens in place p$_2$:

>ancestor (jk, jl),
>ancestor (jk, je),
>ancestor (jl, pr),
>ancestor (jl, my),
>ancestor (my, jn).

Next, transition tr$_2$ fires generating the following tokens as the solution to the goal

>'←ancestor (X, Y).':

>ancestor (jk, pr),
>ancestor (jk, my),
>ancestor (jl, jn).

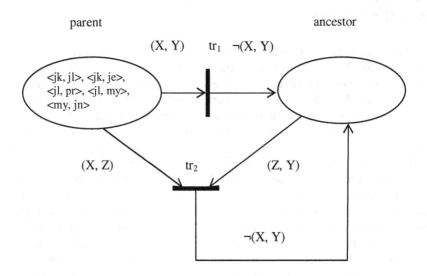

Fig. 6.6: The Petri net approach

Thus only two firing cycles are required to execute the logic program. It may be noted that both top-down approach of PROLOG and the bottom-up approach of LDL requires significant computational time to execute the aforementioned program.

Example 6.8: In this example we consider a practical database problem with respect to a banking system.

Given the following relations:

account (branch-name, account-number, balance)
interest-rate (account-number, percentage-rate)

Suppose the following Datalog rules are given

interest-rate (A, 0) ←account (N, A, B), B < 2000.
interest-rate (A, 5) ←account (N, A, B), B >= 2000.

The aforementioned rules and facts, as given in table 6.4, are mapped onto a Petri net (before firing Fig. 6.7 (a) and after firing Fig 6.7 (b)) and the 'Procedure Automated-reasoning' (vide chapter 3) is invoked to answer the user-made query:

? interest-rate (A, I).

Table 6.4: The 'account' relation.

Branch-name (N)	Account-number (A)	Balance in Rupees (B)
Sodpur	0215	1000
C R Avenue	0305	4500
Park Circus	0201	8500

The machine in turn responds with table 6.5.

Table 6.5: The 'interest-rate' relation.

Account-number (A)	Interest-rate in percentage (I)
0215	0
0305	5
0201	5

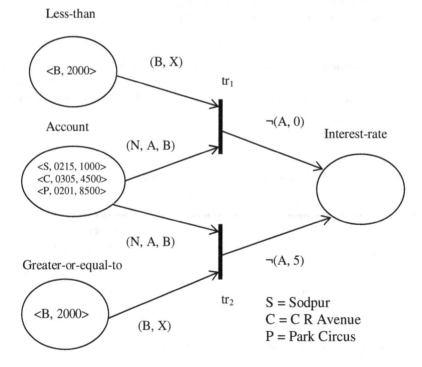

Fig. 6.7 (a): The Petri net before firing

The Petri net after firing is presented in Fig. 6.7 (b)).

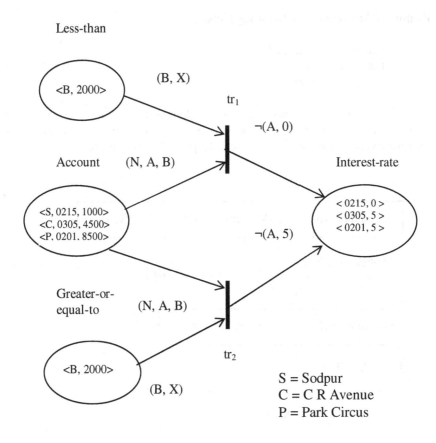

Fig. 6.7 (b): The Petri net after firing

6.7 Scope of Petri Net-based Model in Data Mining

The term Data Mining loosely refers to discovering knowledge from a large volume of data. It has similarity with knowledge discovery in artificial intelligence [3]. There are many methods of data mining. Some of them include statistical techniques, clustering, Bayessian scheme, neural net approach and many others. In recent times, researchers are taking keen interest to automatically extract knowledge from a given set of first order rules and facts. This is well known as *Inductive Logic Programming (ILP)*. In ILP, the well-known resolution theorem is employed in a backward sense. To illustrate the scope of ILP in data mining let us consider example 6.9.

Example 6.9: Consider the following facts:

> Grandfather (janak, lab) ←.
> Father (janak, sita) ←.
> Mother (sita, lab) ←.

We are interested to construct any new knowledge by applying resolution theorem in an inverted manner onto the above clauses. Figure 6.8 presents a schematic view of the knowledge generation process in two discrete steps.

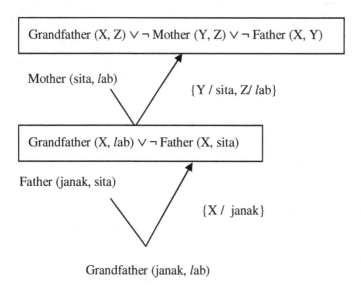

Fig. 6.8: An inverse resolution procedure applied to the given facts finally to derive a knowledge: Grandfather (X, Z) ←Mother (Y, Z) ∧ Father (X, Y)

The same scheme can also be realized using Petri net model with a slight modification in the nomenclature of the Petri net model introduced earlier. A simple scheme illustrating the procedure of knowledge extraction using Petri net is introduced here. The Petri net in the present context needs to be fired in backward direction. In other words given one fact located at the input place of a transition and a resulting inference located at the output place of the transition, we can always generate a rule at the other input place of the same transition. Here the resulting rule is generated by **or**ing the negation of the input fact and the inferred fact at the input place denoted by a box. For generalization, the common value of tokens at the given input and the output place may be replaced by a variable.

Example 6.10: Given the same knowledge base as introduced in example 6.9. We in the present example discuss the scope of Petri nets to extract the same knowledge (vide Fig. 6.9).

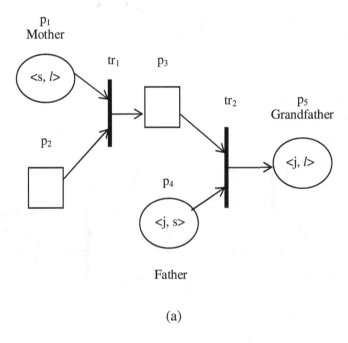

(a)

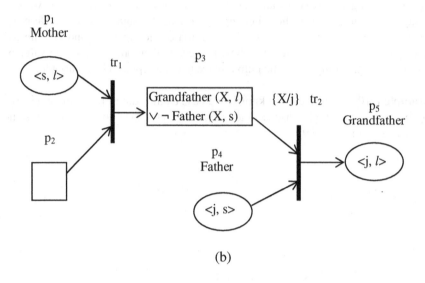

(b)

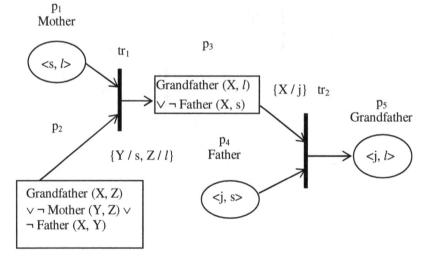

(c)

Fig. 6.9: The order of transition firing to generate the knowledge:
Grandfather (X, Z) ∨ ¬ Mother (Y, Z) ∨ ¬ Father (X, Y).
where s = sita, *l* = *l*ab, j = janak

(a) Initial configuration of the extended Petri net
(b) After firing of transition tr₂
(c) After firing of transition tr₁

Data mining using Petri nets can also be extended for general predicate logic based system. Under this circumstance a transition can have more than one output place and tokens must conform with the respective arc function variables at all input and output places excluding only one input place of the transition. The process of backward transition firing will be continued from the concluding place until no further transition firing is possible. The sequential firing of transitions in a backward sense, supporting the rules of resolution theorem, thus ensures soundness of the derived knowledge.

6.8 Conclusions

Typical database programs are executed on loosely coupled system architecture, where the front-end and the back-end can be easily isolated from each other. Such architectures have several limitations, which can be overcome by realizing database programs on a tightly coupled system. The LDL is one such tightly coupled data language that supports logic programming for data manipulation and management. The LDL system has a number of advantages over other traditional logic program systems, but it is incapable to detect all possible parallelisms in a logic program. Petri net based models for logic programming is introduced in chapter 3 of the book, however, provide a framework for massive parallelism of logic programming and can ensure the execution of all possible parallel resolutions in the logic program. Thus in absence of any constraint on hard ware resources, Petri net based model is the ideal choice for logic program based database machines.

Exercises

1. Consider the following two relations:

    supervises (X, Y) and superior (X, Y).

    The following facts and rules are given in a database.

Cl_1: supervises (f, n) ←.
Cl_2: supervises (f, r) ←.
Cl_3: supervises (f, b) ←.
Cl_4: supervises (e, k) ←.
Cl_5: supervises (e, h) ←.
Cl_6: supervises (a, f) ←.
Cl_7: supervises (a, e) ←.
Cl_8: superior (X, Y) ←supervises (X, Y).
Cl_9: superior (X, Y) ←supervises (X, Z), superior (Z, Y).
Cl_8: subordinate (X, Y) ←superior (Y, X).

Given the query:

$$? \text{ superior } (a, Y) \leftarrow.$$

Construct a top-down traversal tree to answer the query.

[**Hints:** A top-down traversal tree is constructed with the facts and rules given in the database vide Fig. 6.10.

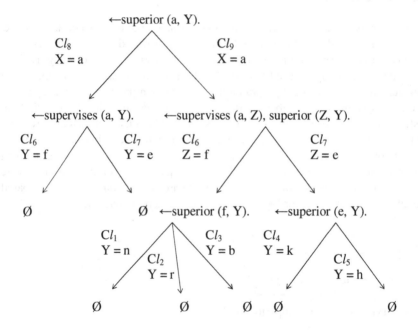

Fig. 6.10: The supervisory-tree based on the given facts

Here, the answers to the query are:

$$\text{superior } (a, f) \leftarrow.$$
$$\text{superior } (a, e) \leftarrow.$$
$$\text{superior } (a, n) \leftarrow.$$
$$\text{superior } (a, r) \leftarrow.$$
$$\text{superior } (a, b) \leftarrow.$$
$$\text{superior } (a, k) \leftarrow.$$
$$\text{superior } (a, h) \leftarrow.$$]

2. Consider the following relations in a banking system:

    account (branch-name, account-number, customer-name, balance),
    depositor (customer-name, account-number),
    borrower (customer-name, loan).

    Construct the following queries using the above relations.

    (a) Identify the account-number whose balance is above Rs 10000 and is not
        enlisted in the borrower relation.

    (b) Determine the account-number whose balance is above Rs 1500 and has
        taken a loan less than Rs 5000.

    (c) Also construct necessary tables to describe the relations and answer the
        aforementioned queries using the tables.

    [**Hints:** Let us abbreviate the attributes of the relations as follows:

                    N for branch-name,
                    A for account-number,
                    C for customer-name,
                    B for balance,
                    L for loan.

    a) account (N, A, C, B) ←B > 10000, **not**-is –borrower (C, L).
    b) account (N, A, C, B) ←B > 1500, borrower (C, L), L < 5000.
    c) Representative Tables 6.6 to 6.8 have been constructed to answer the
        queries listed in parts (a) and (b).

Table 6.6: The 'account' relation.

Branch-name (N)	Account-number (A)	Customer-name (C)	Balance in rupees (B)
Jadavpur	0101	ram	19000
Sodpur	0215	sabi	1000
Maniktala	0102	rai	9000
Park Circus	0322	ali	26500
C R Avenue	0305	sush	4500
Park Circus	0201	madhu	8500
Park Street	0222	john	7500

Table 6.7: The 'depositor' relation.

Customer-name (N)	Account-number (A)
ram	0101
shyam	0200
jadu	0021
madhu	0201
ganga	0120
hari	0221

Table 6.8: The 'borrower' relation.

Customer-name (N)	Loan (L)
ram	4000
shyam	500
jadu	2000
sita	5000
mira	3000
rai	7000
kali	1200

The answer to the first query: account (Park Circus, 0322, ali, 26500).
The answer to the second query: account (Jadavpur, 0101, ram, 19000).]

3.   Consider the following rules and facts:

> Herbivore (X) ←Lives-on-grass (X), Four-footed (X).
> Lives-on-grass (cow) ←.
> Lives-on-grass (buffalo) ←.
> Four-footed (cow) ←.

Using closed-world assumption and negation by failure answer the following three queries:

(a) Query 1: ? Herbivore (cow).
(b) Query 2: ? Four-footed (buffalo).
(c) Query 3: ? Herbivore (buffalo).

[**Hints:** By closed-world assumption we state that ¬Four-footed (buffalo) is true. Consequently answer to part (b) and (c) are false. The answer to part (a), however, is true as the premises to derive Herbivore (cow) are supplied in the database.]

4. Use stratification strategy of LDL to answer query for the given logic program:

Rules: One-way-route (X, Y) ←Route (X, Y), ¬Destination (X).
         Two-way-route (X, Y) ←Route (X, Y).

Query: ? One-way-route (X, Y).

Facts: Route (a, b) ←.
         Route (b, d) ←.
         Route (c, d) ←.
         Route (d, a) ←.
         Destination (a) ←.
         Destination (c) ←.
         Destination (d) ←.

[**Hints:** First identify X, for which destination X is not available in the database. The routes thus generated using the first rule yield one-way-route. The remaining routes are two-way.]

References

1. Aho, A. and Ullman, J., "Universality of Data Retrieval Languages," *proceedings of the POPL conference*, San Antonio Tx, ACM, 1979.
2. Bancilhon, F. and Ramakrishnan, R., "An Amateur's Introduction to Recursive Query Proceeding Strategies," in *proceedings of the ACM SIGMOD International Conference on Management of Data*, 1986.
3. Bry, F., "Query Evaluation in Recursive Databases: Bottom-up and Top-down reconciled," *IEEE Transactions on Knowledge and Data Engineering*, 2, 1990.
4. Ceri, S., Gottlob, G., Tanca, L., *Logic Programming and Databases*, Springer-Verlag, 1990.
5. Chang, C., " On the Evaluation of Queries containing Derived Relations in a Relational Database," in *Advances in Database Theory*, Gallaire, H., Minker, J. and Nicolas, J. (Eds.), vol. 1, Plenum press, 1981.
6. Chimenti, D. et al., "An Overview of the LDL System," *MCC Technical Report # ACA-ST-370-87*, Austin, Tx, November 1987.
7. Chimenti, D., Gamboa, R., Krishnamurthy, R., Naqvi, S., Tsur, S. and Zaniolo, C., "The LDL system Prototype," *IEEE Trans. on Knowledge and Data Engineering*, vol. 2, no. 1, pp. 76-90, March 1990.
8. Elmasri, R. and Navathe, S. B., *Fundamentals of Database Systems*, The Benjamin/Cummings, CA, 1994.

9. Fayyad, U. M., Piatetsky-Shapiro, G., Smyth, P. and Uthurusamy, R. (Eds.), *Advances in Knowledge Discovery and Data Mining*, AAAI Press/The MIT Press, Menlo Park, CA, pp. 117-152, 1996.
10. Gallaire, H., Minker, J. (Eds.), *Logic and Databases*, Plenum press, 1978.
11. Gallaire, H., Minker, J. and Nicolas, J., "Logic and Databases: A Deductive Approach," *ACM Computing Surveys*, 16:2, June1984.
12. Kifer, M. and Lozinskii, E., "A Framework for an Efficient Implementation of Deductive Databases," *proceedings of the sixth Advanced Database Symposium*, Tokyo, Japan, August 1986.
13. Konar, A., *Artificial Intelligence and Soft Computing: Behavioral and Cognitive Modeling of the Human Brain*, CRC Press, Boca Raton, Florida, 1999.
14. Krishnamurthy, R. and Naqvi, S., "Database Updates in Logic Programming, Rev. 1," *MCC Technical Report # ACA-ST-010-88*, Rev. 1, September 1988.
15. Krishnamurthy, R. and Naqvi, S., "Non-Deterministic Choice in Datalog," *proceedings of the 3rd International Conference on Data and Knowledge Bases*, June 27-30, Jerusalem, Israel, 1989.
16. Levesque, H., "The Logic of Incomplete Knowledge Bases," in *On Conceptual Modeling*, Broadie, M., Mylopoulos, J., and Schmidt, J. (Eds.), SpringerVerlag, 1984.
17. Ramkrishnan, R., Srivastava, D. and Sudarshan, S., "{CORAL}: {C}ontrol, {R}elations and {L}ogic," in *proceedings of the International Conference on Very Large Data Bases*, 1992.
18. Ramkrishnan, R., Srivastava, D., Sudarshan, S., and Sheshadri, P., "Implementation of the {CORAL} deductive database system," in *proceedings of the ACM SIGMOD International Conference on Management of Data*, 1993.
19. Reiter, R., "Towards a Logical Reconstruction of Relational Database Theory," in *On Conceptual Modeling*, Broadie, M., Mylopoulos, J., and Schmidt, J. (Eds.), SpringerVerlag, 1984.
20. Silberschatz, A., Korth, H. F. and Sudarshan, S., *Database System Concepts*, McGraw-Hill, Singapore, 1997.
21. Tsur, S., "LDL – A Technology for the Realization of Tightly Coupled Expert Database Systems," *IEEE Expert Magazine*, pp. 41-51, Fall 1988.
22. Ullman, J., *Principles of Database and Knowledge-Base Systems*, vol. 1, Computer Science Press, 1988.
23. Ullman, J., *Principles of Database and Knowledge-Base Systems*, vol. 2, Computer Science Press, 1989.
24. Vielle, L., "Recursive Axioms in Deductive Databases: The Query-Subquery Approach," in *proceedings of the International Conference on Expert Database Systems*, 1986.
25. Vielle, L., "Database Complete Proof Production based on SLD-resolution," in *proceedings of the 4th International Conference on Logic Programming*, 1987.

26. Whang, K. and Navathe, S., "Integrating Expert Systems with Database Management Systems- an Extended Disjunctive Normal Form Approach," in *Information Sciences*, 64, March 1992.
27. Zaniolo, C., "Design and Implementation of a Logic Based Language for Data Intensive Applications," *MCC Technical Report # ACA-ST-199-88*, June 1988.

Appendix A

Simulation of the Proposed Modular Architecture

This appendix provides an outline to the simulation aspects of the proposed architectures using VHDL. The VHDL source codes for all the modules are too large to be included in this appendix. For convenience of the readers, we provide the source code for the matcher circuit only.

A.1 Introduction

The architecture proposed in chapter 4 was simulated in both C and VHDL[1] languages. The C-realization was necessary to verify the functional behavior of the individual modules. The VHDL-realization, on the other hand, provides a detailed implementation of the modules with logic gates. The timing details of the proposed functional architecture was also tested and verified using the VHDL simulation.

The VHDL language benefits the users on the following counts: (i) simplicity in submission of a formal description of the structure to be designed, (ii) the scope of decomposition of a design into sub-designs and (iii) simplicity in establishing interconnections among the sub-designs. The advantage of VHDL simulation lies in testing and verifying the logic implementation without the expense of *hardware prototyping*.

The sub-units, such as Transition History File (THF), Place Token Variable Value Mapper (PTVVM), etc., presented in chapter 4 have been developed under separate *projects*. The different entities of each sub-unit were coded either by *structures* or *functional behaviors* in one or two *modules* under the specific project. The port connections among the entities were accomplished through *signals* in a different module, referred to as the top-level module, containing the top-level entity. To test the correctness of realization, the top-level entity that represents a sub-unit of the architecture was simulated. The simulation was performed on a module called the *test-bench*. The test-bench was excited with

---

[1] VHDL is a language for describing digital electronic systems. It arose out of the United States Government's **V**ery **H**igh-**S**peed **I**ntegrated **C**ircuits (VHSIC) Program. The **VHSIC H**ardware **D**escription **L**anguage (VHDL) was developed for testing and verifying the structure and function of Integrated Circuits.

sample inputs and its logical output waveforms were verified through VHDL simulation.

The VHDL implementation of one typical circuit (the Matcher) is presented below for convenience. The complete code for the VHDL-implementation is available in one recent undergraduate thesis [1], prepared under the active co-ordination by the authors.

A.2 VHDL Code for Different Entities in Matcher

```
library ieee;

use ieee.std_logic_1164.all;

        ----------ENTITY TWO INPUT AND GATE----------
entity and_2_gate is
   port(A:in std_ulogic;
        B:in std_ulogic;
        C:out std_ulogic);

end and_2_gate;

architecture behaviour of and_2_gate is
 begin
   C <= A and B after 10ns;
 end behaviour;

        ----------ENTITY FIVE INPUT AND GATE----------
use ieee.std_logic_1164.all;

entity and_5_gate is
 port( A : in std_ulogic;
       B : in std_ulogic;
       C : in std_ulogic;
       D : in std_ulogic;
       E : in std_ulogic;
       F : out std_ulogic );
end and_5_gate;

architecture behaviour of and_5_gate is
 begin
       F <= A and B and C and D and E after 10 ns;
 end behaviour;
```

```
----------ENTITY XNOR GATE----------

use ieee.std_logic_1164.all;

entity not_xor is
 port ( I1 : in std_ulogic_vector( 4 downto 0);

         I2 : in std_ulogic_vector( 4 downto 0);
         X1 : out std_ulogic_vector( 4 downto 0));
 end not_xor;


architecture behaviour of not_xor is
 signal temp_X1 : std_ulogic_vector( 4 downto 0);
 begin
   temp_X1 <= I1 xor I2;
   X1 <= not temp_X1;
 end;


----------ENTITY AND TREE----------

use ieee.std_logic_1164.all;

entity AND_TREE is
 port ( TI1 : in std_ulogic_vector (4 downto 0);
        TI2 : in std_ulogic_vector (4 downto 0);
        TI3 : in std_ulogic_vector (4 downto 0);
        TI4 : in std_ulogic_vector (4 downto 0);
        TI5 : in std_ulogic_vector (4 downto 0);
        TO1 : out std_ulogic);
 end AND_TREE;

architecture structure of AND_TREE is
 component and_2_gate
   port(A:in std_ulogic;
        B:in std_ulogic;
        C:out std_ulogic);
 end component;

 component and_5_gate
 port( A : in std_ulogic;
       B : in std_ulogic;
       C : in std_ulogic;
```

```
      D : in std_ulogic;
      E : in std_ulogic;
      F : out std_ulogic  );
   end component;


component not_xor
  port ( I1 : in std_ulogic_vector( 4 downto 0);
         I2 : in std_ulogic_vector( 4 downto 0);
         X1 : out std_ulogic_vector( 4 downto 0));
  end component;

  signal Q1,Q2,Q3,Q4:std_ulogic_vector ( 4 downto 0);
  signal TEMP_0_1,TEMP_0_2,TEMP_0_3,TEMP_0_4 :
std_ulogic;
  signal TEMP_1_1,TEMP_1_2,TEMP_1_3,TEMP_2_1,TEMP_2_2:
std_ulogic;

  begin

  INST1_not_xor : not_xor port map ( I1 => TI1 , I2 =>
TI2 , X1 => Q1 );
  INST2_not_xor : not_xor port map ( I1 => TI2 , I2 =>
TI3 , X1 => Q2 );
  INST3_not_xor : not_xor port map ( I1 => TI3 , I2 =>
TI4 , X1 => Q3 );
  INST4_not_xor : not_xor port map ( I1 => TI4 , I2 =>
TI5 , X1 => Q4 );
  INST1_and_5_gate:and_5_gate port
map(A=>Q1(0),B=>Q1(1),C=>Q1(2),D=>Q1(3),E=>Q1(4),F=>TEM
P_0_1);
  INST2_and_5_gate:and_5_gate
                        port map (
A=>Q2(0),B=>Q2(1),C=>Q2(2),D=>Q2(3),E=>Q2(4),F=>TEMP_0_
2);
  INST3_and_5_gate:and_5_gate
                        port map (
A=>Q3(0),B=>Q3(1),C=>Q3(2),D=>Q3(3),E=>Q3(4),F=>TEMP_0_
3);
  INST4_and_5_gate:and_5_gate
                        port map (
A=>Q4(0),B=>Q4(1),C=>Q4(2),D=>Q4(3),E=>Q4(4),F=>TEMP_0_
4);
```

```
 INST1_and_2_gate:and_2_gate port map(
A=>TEMP_0_1,B=>TEMP_0_2,C=>TEMP_1_1);
 INST2_and_2_gate:and_2_gate port map(
A=>TEMP_0_2,B=>TEMP_0_3,C=>TEMP_1_2);

 INST3_and_2_gate:and_2_gate port map(
A=>TEMP_0_3,B=>TEMP_0_4,C=>TEMP_1_3);
 INST4_and_2_gate:and_2_gate port map(
A=>TEMP_1_1,B=>TEMP_1_2,C=>TEMP_2_1);
 INST5_and_2_gate:and_2_gate port map(
A=>TEMP_1_2,B=>TEMP_1_3,C=>TEMP_2_2);
 INST6_and_2_gate:and_2_gate port map(
A=>TEMP_2_1,B=>TEMP_2_2,C=>TO1);

 end structure;
```

----------**ENTITY REGISTER MATCH**----------

```
 library ieee;
use ieee.std_logic_1164.all;

 entity register_match is
   port( A1 :in std_ulogic_vector (4 downto 0);
         A2 :in std_ulogic_vector (4 downto 0);
         A3 :in std_ulogic_vector (4 downto 0);
         A4 :in std_ulogic_vector (4 downto 0);
         A5 :in std_ulogic_vector (4 downto 0);
         C1 :out std_ulogic_vector (4 downto 0);
         C2 :out std_ulogic_vector (4 downto 0);
         C3 :out std_ulogic_vector (4 downto 0);
         C4 :out std_ulogic_vector (4 downto 0);
         C5 :out std_ulogic_vector (4 downto 0));
 end register_match;
architecture behaviour of register_match is

   signal
reg_val1,reg_val2,reg_val3,reg_val4,reg_val5:std_ulogic
_vector (4     downto 0);
   signal AND1,AND2,AND3,AND4,AND5:std_ulogic;
   signal check:std_ulogic := 'X';
     begin

            reg_val1<= A1;
```

```
          reg_val2<= A2;
          reg_val3<= A3;
          reg_val4<= A4;

      reg_val5<= A5;

   LOADPROCESS:process

   begin


     AND1 <= reg_val1(0) and reg_val1(1) and
reg_val1(2) and reg_val1(3) and reg_val1(4);

     AND2 <= reg_val2(0) and reg_val2(1) and
reg_val2(2) and reg_val2(3) and reg_val2(4);-- after 10
ns;
     wait for 10 ns;
     if ( AND1 = '1') then
           reg_val1 <= reg_val2;
      end if;
     if ( AND2 = '1') then
           reg_val2 <= reg_val1;
      end if;
     AND2 <= reg_val2(0) and reg_val2(1) and
reg_val2(2) and reg_val2(3) and reg_val2(4) ;--after 10
ns;
     AND3 <= reg_val3(0) and reg_val3(1) and
reg_val3(2) and reg_val3(3) and reg_val3(4) ;--after 10
ns;
     wait for 10 ns;

     if ( AND2 = '1') then
           reg_val2 <= reg_val3;
      end if;
     if ( AND3 = '1') then
           reg_val3 <= reg_val2;
      end if;

     AND3 <= reg_val3(0) and reg_val3(1) and
reg_val3(2) and reg_val3(3) and reg_val3(4) ;--after 10
ns;
```

```
        AND4 <= reg_val4(0) and reg_val4(1) and
reg_val4(2) and reg_val4(3) and reg_val4(4)  ;--after 10
ns;
        wait for 10 ns;

        if ( AND3 = '1') then
                reg_val3 <= reg_val4;
         end if;
        if ( AND4 = '1') then
                reg_val4 <= reg_val3;
         end if;

        AND4 <= reg_val4(0) and reg_val4(1) and
reg_val4(2) and reg_val4(3) and reg_val4(4)  ;--after 10
ns;
        AND5 <= reg_val5(0) and reg_val5(1) and
reg_val5(2) and reg_val5(3) and reg_val5(4)  ;--after 10
ns;
        wait for 10 ns;

        if ( AND4 = '1') then
                reg_val4 <= reg_val5;
         end if;
        if ( AND5 = '1') then
                reg_val5 <= reg_val4;
         end if;
                check <= '1';
         wait for 10 ns;
        if ( check ='1') then
        C1 <= reg_val1;
        C2 <= reg_val2;
        C3 <= reg_val3;
        C4 <= reg_val4;
        C5 <= reg_val5;
        end if;
      end process;
  end behaviour;
```

A.3 VHDL Code to Realize the Top Level Architecture of Matcher

```
library ieee;

use ieee.std_logic_1164.all;
use work.all;
```

----------**ENTITY TOP LEVEL ARCHITECTURE**----------

```
entity TOP is
        port(X1 : in std_ulogic_vector ( 4 downto 0 );
             X2 : in std_ulogic_vector ( 4 downto 0 );
             X3 : in std_ulogic_vector ( 4 downto 0 );
             X4 : in std_ulogic_vector ( 4 downto 0 );
             X5 : in std_ulogic_vector ( 4 downto 0 );
        OUTFINAL: out std_ulogic_vector ( 4 downto 0 );
          OUTBIT: out std_ulogic);
end TOP    ;


architecture STRUCTURE of TOP is

component register_match
        port( A1 :in std_ulogic_vector (4 downto 0);
              A2 :in std_ulogic_vector (4 downto 0);
              A3 :in std_ulogic_vector (4 downto 0);
              A4 :in std_ulogic_vector (4 downto 0);
              A5 :in std_ulogic_vector (4 downto 0);
              C1 :out std_ulogic_vector (4 downto 0);
              C2 :out std_ulogic_vector (4 downto 0);
              C3 :out std_ulogic_vector (4 downto 0);
              C4 :out std_ulogic_vector (4 downto 0);
              C5 :out std_ulogic_vector (4 downto 0));
end component;

component AND_TREE
    port ( TI1 : in std_ulogic_vector (4 downto 0):=
"XXXXX";
           TI2 : in std_ulogic_vector (4 downto 0):=
"XXXXX";
           TI3 : in std_ulogic_vector (4 downto 0):=
"XXXXX";
           TI4 : in std_ulogic_vector (4 downto 0):=
"XXXXX";
           TI5 : in std_ulogic_vector (4 downto 0):=
"XXXXX";
           TO1 : out std_ulogic);
 end component;

signal t_TI1,t_TI2,t_TI3,t_TI4,t_TI5 :
std_ulogic_vector (4 downto 0);
```

```
SIGNAL t_OUTBIT : std_ulogic ;

begin

   INST_REGISTER_MATCH:register_match
                                   port map(A1 => X1,
                                            A2 => X2,
                                            A3 => X3,
                                            A4 => X4,
                                            A5 => X5,
                                            C1 => t_TI1,
                                            C2 => t_TI2,
                                            C3 => t_TI3,
                                            C4 => t_TI4,
                                            C5 => t_TI5);

   INST_AND_TREE: AND_TREE
                                   port map( TI1 => t_TI1,
                                             TI2 => t_TI2,
                                             TI3 => t_TI3,
                                             TI4 => t_TI4,
                                             TI5 => t_TI5,
                                          TO1 => t_OUTBIT);

   INST : process(t_outbit)
   begin
    if (t_OUTBIT = '1' ) then
        OUTBIT <= t_OUTBIT ;
    elsif (t_OUTBIT = '0') then
             OUTBIT <= t_OUTBIT ;
    else
          OUTBIT <= 'X';
       end if;
    if ( t_OUTBIT = '1' ) then
       OUTFINAL <= t_TI5;
    elsif ( t_OUTBIT = '0' ) then
       OUTFINAL <= "XXXXX";
    end if;
    end process;
end STRUCTURE;
```

A.4 VHDL Code of Testbench to Simulate the Matcher

```
library ieee;

use ieee.std_logic_1164.all;
use std.textio.all;
use work.all;

        --------ENTITY TESTBENCH OF MATCHER--------
entity test is
end test;

architecture matcher_stimulus of test is

 component TOP
     port( X1 : in std_ulogic_vector ( 4 downto 0 );
           X2 : in std_ulogic_vector ( 4 downto 0 );
           X3 : in std_ulogic_vector ( 4 downto 0 );
           X4 : in std_ulogic_vector ( 4 downto 0 );
           X5 : in std_ulogic_vector ( 4 downto 0 );
       OUTFINAL : out std_ulogic_vector ( 4 downto 0);
         OUTBIT: out std_ulogic);
 end component;

 signal T_TI1X,T_TI2X,T_TI3X,T_TI4X,T_TI5X,MATCHOUTX:
std_ulogic_vector (4 downto 0);
 signal RESULTX: std_ulogic;
 signal T_TI1Y,T_TI2Y,T_TI3Y,T_TI4Y,T_TI5Y,MATCHOUTY:
std_ulogic_vector (4 downto 0);
 signal RESULTY: std_ulogic;
 signal T_TI1Z,T_TI2Z,T_TI3Z,T_TI4Z,T_TI5Z,MATCHOUTZ:
std_ulogic_vector (4 downto 0);
 signal RESULTZ: std_ulogic;
 signal ANDMATCH : std_ulogic;

begin

  INST1_TOPX :          TOP port map(X1 => T_TI1X,
                                     X2 => T_TI2X,
                                     X3 => T_TI3X,
                                     X4 => T_TI4X,
                                     X5 => T_TI5X,
                               OUTFINAL => MATCHOUTX,
                                 OUTBIT => RESULTX);
```

```
INST1_TOPY :                 TOP port map(X1 => T_TI1Y,
                                          X2 => T_TI2Y,
                                          X3 => T_TI3Y,
                                          X4 => T_TI4Y,
                                          X5 => T_TI5Y,
                                  OUTFINAL => MATCHOUTY,
                                    OUTBIT => RESULTY);

INST1_TOPZ :                 TOP port map(X1 => T_TI1Z,
                                          X2 => T_TI2Z,
                                          X3 => T_TI3Z,
                                          X4 => T_TI4Z,
                                          X5 => T_TI5Z,
                                  OUTFINAL => MATCHOUTZ,
                                    OUTBIT => RESULTZ);

ANDMATCH <= RESULTX and RESULTY and RESULTZ;
MATCHER_PROCESS: process
     Begin

                   T_TI1X <= "10001";
                   T_TI2X <= "11111";
                   T_TI3X <= "10001";
                   T_TI4X <= "11111";
                   T_TI5X <= "10001";
                   T_TI1Y <= "10001";
                   T_TI2Y <= "11111";
                   T_TI3Y <= "10001";
                   T_TI4Y <= "11111";
                   T_TI5Y <= "10001";
                   T_TI1Z <= "10001";
                   T_TI2Z <= "11111";
                   T_TI3Z <= "10001";
                   T_TI4Z <= "11111";
                   T_TI5Z <= "10001";

                   wait for 200 ns;
                   T_TI1X <= "10101";
                   T_TI2X <= "10001";
                   T_TI3X <= "10001";
                   T_TI4X <= "11111";
                   T_TI5X <= "11111";
                   T_TI1Y <= "10101";
                   T_TI2Y <= "10001";
```

```
          T_TI3Y <= "10001";
          T_TI4Y <= "11111";
          T_TI5Y <= "11111";
          T_TI1Z <= "10101";
          T_TI2Z <= "10001";
          T_TI3Z <= "10001";
          T_TI4Z <= "11111";
          T_TI5Z <= "11111";

wait for 200 ns;
          T_TI1X <= "10001";
          T_TI2X <= "10001";
          T_TI3X <= "10011";
          T_TI4X <= "11111";
          T_TI5X <= "11111";
          T_TI1Y <= "10001";
          T_TI2Y <= "10001";
          T_TI3Y <= "10011";
          T_TI4Y <= "11111";
          T_TI5Y <= "11111";
          T_TI1Z <= "10001";
          T_TI2Z <= "10001";
          T_TI3Z <= "10011";
          T_TI4Z <= "11111";
          T_TI5Z <= "11111";

          wait for 200 ns;
     end process;
end matcher_stimulus ;
```

Reference

1. Mukherjee, R. and Mukhopadhyay, S., *VHDL-implementation for a parallel architecture for logic programming*, undergraduate thesis, Jadavpur University, 2000.

Appendix B

Open-ended Problems for Dissertation Works

This appendix provides some open-ended problems of common interest and is recommended for extension by graduate students, pursuing their research in the area of parallel and distributed logic programming. Research directions to solve these problems are also provided to motivate young researchers to undertake these problems for their research.

B.1 Problem 1: The Diagnosis Problem

There exist two alternative approaches to solve a diagnosis problem. The first approach, well known as *model-based approach* [1] attempts to develop a forward (simulation) model of the system to be diagnosed, and then employs a diagnostic algorithm on this model to determine the abnormal behavior in the systems, if any. The cause of abnormality is diagnosed at a later stage to identify the system components responsible for the abnormality.

Unlike the model-based approach, an alternative approach to handle the diagnosis problem is to construct a set of diagnostic rules, depicting the knowledge of a skilled engineer, to correctly detect the defective components in a system, if any. The *rule-based* paradigm can directly be realized on Petri nets for automated reasoning in a diagnosis problem.

In this section, we illustrate the scope of Petri net models in solving diagnosis problems [2]. The diagnostic rules for a full wave two-diode rectifier system are given below.

Rule 1: Transformer-output (0V), Open (one-half-of-secondary-coil)
←Defective (transformer), Primary (230V).

Rule 2: Defective (one-diode) ←Defective (rectifier).

Rule 3: Defective (two-diodes) ←Defective (rectifier).

Rule 4: Rectifier-output (0V) ←Transformer-output (0V).

Rule 5: Rectifier-output (6V) ←Open (one-half-of-secondary-coil).

Rule 6: Rectifier-output (6V) ←Defective (one-diode).

Rule   7:   Rectifier-output (0V) ←Defective (two-diodes).

The diagnostic knowledge given by Rules 1 to 7 can be represented in a structured manner by a Petri net model shown in Fig. B.1.

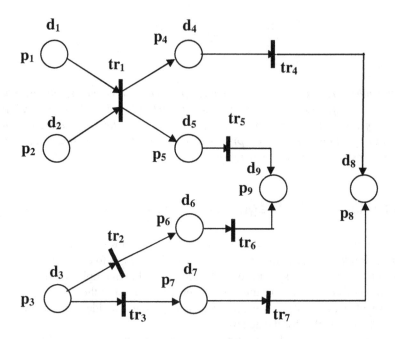

d_1 = Defective (transformer), d_2 = Primary (230V), d_3 = Defective (rectifier), d_4 = Transformer-output (0V), d_5 = Open (one-half-of-secondary-coil), d_6 = Defective (one-diode), d_7 = Defective (two-diodes), d_8 = Rectifier-output (0V), d_9 = Rectifier-output (6V)

Fig. B.1: A Petri net representing diagnostic knowledge of a two-diode full wave rectifier

Show various steps of forward and backward chaining when the tokens are given in the places p_2, p_5 and p_8, and hence comment on the solution to the problem.

[**Hints:** First fire transitions tr_4 and tr_7 in parallel both and in backward manner. Then fire transitions tr_3 and tr_1 both in backward manner. Thus we obtain the truth value of the predicates: Defective-rectifier (place p_3) and Defective-transformer (place p_1).]

B.2 Problem 2: A Matrix Approach for Petri Net Representation

The classical models of *modus ponens* informally described by the following two rules:

$$q \leftarrow p.$$

$$p \leftarrow.$$

$$\overline{q \leftarrow.}$$

where p and q denotes two propositions. We can represent the same concept by vector-matrix approach as follows.

Let **R** (p, q) denotes a binary implication relation between the propositions p and q. Assuming 'q ←p.' to be equivalent to ¬p ∨ q, where '¬' and '∨' have usual meanings, we can represent **R** (p, q) by

$$\mathbf{R}\ (p, q) = \quad \begin{array}{c} q \\ \diagdown \\ p \end{array} \begin{array}{c|cc} & 0 & 1 \\ \hline 0 & 1 & 1 \\ 1 & 0 & 1 \end{array}$$

where the elements of the matrix are evaluated by the formula ¬p ∨ q.

Suppose, p is true, and we need to evaluate the truth value of q by vector-matrix approach, satisfying *modus ponens*. Since we do not know the truth value of q beforehand. We consider it as don't care (d). Thus the initial value of [p q] is given by

$$\begin{pmatrix} p & q \\ 1 & d \end{pmatrix}$$

The inferred value of q now is obtained by taking max-min composition [3] of the above vector with the matrix **R** (p, q).

The inferred value of [p q] is thus given by

$$\begin{array}{cc} p & q \\ \begin{bmatrix} 1 & d \end{bmatrix} \end{array} o \qquad \begin{array}{c} p\downarrow \\ \\ \end{array} \begin{array}{ccc} & q\rightarrow & \\ 0 & 1 \\ \begin{pmatrix} 1 & 1 \\ 0 & 1 \end{pmatrix} \\ 1 & \end{array}$$

$$= \begin{array}{cc} p & q \\ [1 & 1], \end{array}$$

which yields the inferred value of q to be 1.

The same principle can be extended for Petri net models by using two binary matrices **P** and **Q** where **P** denotes a *connectivity from transition to places* and **Q** denotes a *connectivity from places to transitions* respectively.

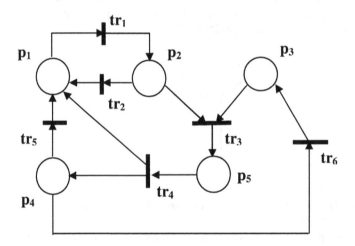

Fig. B.2: A Petri net to illustrate the construction of **P** and **Q** matrices

For illustration, we consider an arbitrary Petri net (Fig. B.2) and evaluate its corresponding **P** and **Q** matrices for the convenience of the readers.

P =

To places \ From trans.	tr_1	tr_2	tr_3	tr_4	tr_5	tr_6
p_1	0	1	0	1	1	0
p_2	1	0	0	0	0	0
p_3	0	0	0	0	0	1
p_4	0	0	0	1	0	0
p_5	0	0	1	0	0	0

Q =

To trans. \ From places	p_1	p_2	p_3	p_4	p_5
tr_1	1	0	0	0	0
tr_2	0	1	0	0	0
tr_3	0	1	1	0	0
tr_4	0	0	0	0	1
tr_5	0	0	0	1	0
tr_6	0	0	0	1	0

Since the *modus ponens* of propositional logic is similar with *forward firing* of a Petri net, we can evaluate the next token in each place when the current token of their predecessors are given.

One general rule for computing binary tokens at the places is given by [2]

$$\mathbf{N}(t+1) = \mathbf{P} \circ [\mathbf{Q} \circ \mathbf{N}^c(t)]^c \qquad (B.1)$$

where $\mathbf{N}(t) = [n_1(t)\ n_2(t).....n_m(t)\ ]$ denotes the tokens (truth/falsehood) $n_1(t)$, $n_2(t)$, ..., $n_m(t)$ at respective places $p_1, p_2, ..., p_m$ at time t. The 'c' above a vector denotes its binary complement.

Graduate students are advised to construct a forward chaining model of Petri net and verify that the given equation (B.1) supports *modus ponens* in a transitive sense for a sequence of forward firable transitions.

Example B.1: Consider a Petri net model as shown in Fig. B.3. Given the $\mathbf{N}(0)$ vector, determine $\mathbf{N}(1)$ and $\mathbf{N}(2)$ and notice the changes in the places p_3 and p_5 in two successive iterations of computing $\mathbf{N}$.

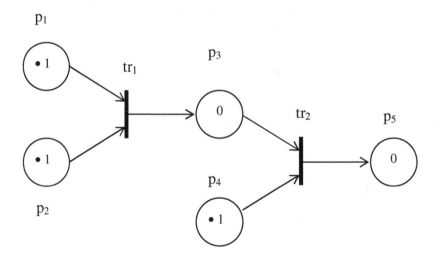

Fig. B.3: An illustrative Petri net

$$n_1 \quad n_2 \quad n_3 \quad n_4 \quad n_5$$
$$\mathbf{N}(0) = [1 \quad 1 \quad 0 \quad 1 \quad 0]$$

From transitions

	tr_1	tr_2
To places		
P_1	0	0
P_2	0	0
p_3	1	0
p_4	0	0
p_5	0	1

$\mathbf{P} =$

From places

	p_1	p_2	p_3	p_4	p_5
To transitions					
tr_1	1	1	0	0	0
tr_2	0	0	1	1	0

$\mathbf{Q} =$

$$\mathbf{N}(1) = \mathbf{P} \circ (\mathbf{Q} \circ \mathbf{N}^C(0))^C$$

$$= \begin{bmatrix} 0 & 0 \\ 0 & 0 \\ 1 & 0 \\ 0 & 0 \\ 0 & 1 \end{bmatrix} \circ \left(\begin{bmatrix} 1 & 1 & 0 & 0 & 0 \\ 0 & 0 & 1 & 1 & 0 \end{bmatrix} \circ \begin{bmatrix} 1 \\ 1 \\ 0 \\ 1 \\ 0 \end{bmatrix}^{C} \right)^{C}$$

$$= \begin{bmatrix} 0 & 0 \\ 0 & 0 \\ 1 & 0 \\ 0 & 0 \\ 0 & 1 \end{bmatrix} \circ \left(\begin{bmatrix} 1 & 1 & 0 & 0 & 0 \\ 0 & 0 & 1 & 1 & 0 \end{bmatrix} \circ \begin{bmatrix} 0 \\ 0 \\ 1 \\ 0 \\ 1 \end{bmatrix} \right)^{C}$$

$$= \begin{bmatrix} 0 & 0 \\ 0 & 0 \\ 1 & 0 \\ 0 & 0 \\ 0 & 1 \end{bmatrix} \circ \begin{bmatrix} 0 \\ 1 \end{bmatrix}^{C}$$

$$= \begin{bmatrix} 0 & 0 \\ 0 & 0 \\ 1 & 0 \\ 0 & 0 \\ 0 & 1 \end{bmatrix} \circ \begin{bmatrix} 1 \\ 0 \end{bmatrix}$$

$$= \begin{bmatrix} 0 & 0 & 1 & 0 & 0 \end{bmatrix}^{T}$$

which indicates that firing of transition tr_1 generates a token at place p_3. In the second iteration, we can generate token at place p_5. This can be accomplished by

$$N(2) = P \text{ o } (Q \text{ o } N^C(1))^C$$

$$= \begin{bmatrix} 0 & 0 \\ 0 & 0 \\ 1 & 0 \\ 0 & 0 \\ 0 & 1 \end{bmatrix} \text{ o } \left(\begin{bmatrix} 1 & 1 & 0 & 0 & 0 \\ 0 & 0 & 1 & 1 & 0 \end{bmatrix} \text{ o } \begin{bmatrix} 0 \\ 0 \\ 1 \\ 0 \\ 0 \end{bmatrix}^C \right)^C$$

$$= \begin{bmatrix} 0 & 0 \\ 0 & 0 \\ 1 & 0 \\ 0 & 0 \\ 0 & 1 \end{bmatrix} \text{ o } \left(\begin{bmatrix} 1 & 1 & 0 & 0 & 0 \\ 0 & 0 & 1 & 1 & 0 \end{bmatrix} \text{ o } \begin{bmatrix} 1 \\ 1 \\ 0 \\ 1 \\ 1 \end{bmatrix} \right)^C$$

$$= \begin{bmatrix} 0 & 0 \\ 0 & 0 \\ 1 & 0 \\ 0 & 0 \\ 0 & 1 \end{bmatrix} \circ \begin{bmatrix} 1 \\ 1 \end{bmatrix}^{\mathrm{C}}$$

$$= \begin{bmatrix} 0 & 0 \\ 0 & 0 \\ 1 & 0 \\ 0 & 0 \\ 0 & 1 \end{bmatrix} \circ \begin{bmatrix} 0 \\ 0 \end{bmatrix}$$

$$= \begin{bmatrix} 0 & 0 & 0 & 0 & 0 \end{bmatrix}^{\mathrm{T}}$$

We are afraid! What is this?

The aforementioned experiment shows that after the second firing, the token at place p_5 is zero, but by *modus ponens* we should expect it to be one. The aforementioned problem occurs as the starting places of the network such as p_1, p_2 and p_4 cannot hold the tokens forever. In fact they loose their tokens only after corresponding transition firing. In order to restore the tokens at the starting places even after transition firing, we have to provide self-loops around the starting places through *virtual transitions* [1]. Fig. B.4 provides the corresponding network of Fig. B.3 with virtual transitions.

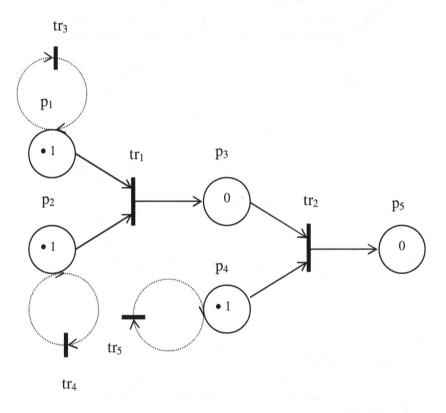

Fig. B.4: Modified Fig. B.3 with virtual transitions around places p_1, p_2 and p_4

The **P** and **Q** matrices for the given Fig. B.4 are changed as follws

P =

	From transitions				
To places	tr_1	tr_2	tr_3	tr_4	tr_5
p_1	0	0	1	0	0
p_2	0	0	0	1	0
p_3	1	0	0	0	0
p_4	0	0	0	0	1
p_5	0	1	0	0	0

Q =

	From places				
To transitions	p_1	p_2	p_3	p_4	p_5
tr_1	1	1	0	0	0
tr_2	0	0	1	1	0
tr_3	1	0	0	0	0
tr_4	0	1	0	0	0
tr_5	0	0	0	1	0

$$\begin{array}{ccccc} n_1 & n_2 & n_3 & n_4 & n_5 \end{array}$$
$$N(0) = [\ \ 1 \quad 1 \quad 0 \quad 1 \quad 0\ \ ]$$

Here, $N(1) = P \circ (Q \circ N^C(0))^C$

$$\begin{array}{ccccc} n_1 & n_2 & n_3 & n_4 & n_5 \end{array}$$
$$= [\ \ 1 \quad 1 \quad 1 \quad 1 \quad 0\ \ ]^T$$

$$N(2) = P \circ (Q \circ N^C(1))^C$$

$$\begin{array}{ccccc} n_1 & n_2 & n_3 & n_4 & n_5 \end{array}$$
$$= [\ \ 1 \quad 1 \quad 1 \quad 1 \quad 1\ \ ]^T$$

which indicates that the new tokens at places p_3 and p_5 after firing of two transitions are one. Further, places p_1, p_2 and p_4 hold their tokens forever, without being hampered by transition firing.

Exercises

1. Extend the aforementioned idea for backward chaining using classical *modus tollens* and combine both forward and backward chaining on Petri net models for reasoning in a logic program.

2. Assuming that the tokens may be non-binary with real values in $[0, 1]$, use equation (B.1) for generating fuzzy inferences in a cycle-free Petri net.

3. Let P^{-1} and Q^{-1} be the inverses [4] of matrices P and Q with respect to max-min composition operation. Assuming that $P^{-1} = P^T$ and $Q^{-1} = Q^T$, we obtain a backward reasoning formalism as follows:

$$N(t+1) = P \circ (Q \circ N^C(t))^C$$
$$\Rightarrow P^{-1} \circ N(t+1) = (Q \circ N^C(t))^C$$
$$\Rightarrow (P^{-1} \circ N(t+1))^C = Q \circ N^C(t)$$
$$\Rightarrow Q \circ N^C(t) = (P^{-1} \circ N(t+1))^C$$

$$\Rightarrow \quad \mathbf{N}^{C}(t) = \mathbf{Q}^{-1} \circ (\mathbf{P}^{-1} \circ \mathbf{N}(t+1))^{C}$$

$$\Rightarrow \quad \mathbf{N}(t) = [\mathbf{Q}^{-1} \circ (\mathbf{P}^{-1} \circ \mathbf{N}(t+1))^{C}]^{C} \qquad \text{(B.2)}$$

Given the token vector $\mathbf{N}(t+1)$, we can obtain $\mathbf{N}(t)$ by using $\mathbf{Q}^{-1} = \mathbf{Q}^{T}$ and $\mathbf{P}^{-1} = \mathbf{P}^{T}$.

Construct a Petri net without cycles (loops) and submit token vector $\mathbf{N}(0)$. Consider self-loop around the starting places through virtual transitions. Now, construct $\mathbf{P}$ and $\mathbf{Q}$ matrices, make several forward passes by iteratively updating equation (B.1) until $\mathbf{N}(t+1) = \mathbf{N}(t)$ at some time $t = t^{*}$.

Now use equation (B.2) to retrieve $\mathbf{N}(0)$ by backward computation of $\mathbf{N}$ vector. Check whether the computed $\mathbf{N}(0)$ is same as the submitted $\mathbf{N}(0)$.

References

1. Hamscher,. W., Console, L., de Kleer, J., *Readings in Model-based Diagnosis,* Morgan-Kaufmann, CA, 1992.
2. Konar, A., *Artificial Intelligence and Soft Computing– Behavioral and Cognitive Modeling of the Human Brain*, CRC Press, Boca Raton, Fl, 2000.
3. Konar, A., *Computational Intelligence: Principles, Techniques and Applications*, Springer, Heidelberg, 2005.
4. Saha, P. and Konar, A., "A heuristic algorithm for computing the max-min inverse fuzzy relation," *International Journal of Approximate Reasoning*, vol. 20, pp. 131-147, 2002.

Index

About the Authors

Alakananda Bhattacharya is currently a senior research associate in a U.G.C.(University Grants Commission)-sponsored project on *Artificial Intelligence applied to Imaging and Robotics*, housed in the department of Electronics and Telecommunications Engineering, Jadavpur University, Calcutta, India. She received her Ph. D. degree on Artificial Intelligence in the sub area of *Parallel Architecture for Logic Programming* in 2002 from the same university. Alakananda has published a number of papers in international journal and conferences in the area of logic programming, database systems and parallel and distributed computing.

The work presented in this book is an extension of Alakananda's research work leading to her Ph. D. degree. It was an extensive work, undertaken over a large period for around ten years to complete the theoretical formalizations, verification and validation of the proposed architecture, and the construction of the compiler for the Datalog type programs.

Alakananda provides peer review for journals and conferences in her field. Her current research interest includes *data* mining by inductive logic programming.

Amit Konar is presently a Professor in the Department of Electronics and Tele-Communication Engineering, Faculty of Engineering and Technology, Jadavpur University, Calcutta, India, and Joint Coordinator, Center for Cognitive Science, Jadavpur University. He is the founding Coordinator of the M.Tech program in Intelligent Automation and Robotics, offered as part-time course to engineering graduates, working in different industries around Calcutta. Dr. Konar has been teaching and carrying out research work at this University for the past 20 years. Jadavpur University is one of the top three universities in India. The faculty of Engineering and Technology, with more than 100 externally funded research projects, earned recognition as a "Center of excellence" from the Government of India, and the university got the "Five Star" rating. The doctoral, master's, and bachelor's programs offered by the university are acknowledged as one of the very best in the country. Books by professors of the university are used as graduate-level texts in Asian, European and American universities. The university runs collaborative programs with European and Canadian universities. Amartya Sen, the 1998 Nobel Laureate in Economics, taught at Jadavpur University for some time.

Dr. Konar's research areas include the study of artificial intelligence algorithms and their applications to the entire domain of Electrical Engineering and Computer Science. Specifically, he worked on logic programming, fuzzy sets and logic, neuro-computing, genetic algorithms, Dempster-Shafer theory and Kalman filtering, and applied the principles of computational intelligence in image understanding, control engineering, VLSI CAD, mobile robotics, bio-informatics and mobile communication systems.

Dr. Konar has published over hundred research papers and several books & invited book chapters on various aspects of computer science and control engineering. His books/ book chapters have been (are in the process of being) published from top publishing houses such as Springer-Verlag, CRC Press, Physica-Verlag, Academic Press, Kluwer Academic Press and Prentice-Hall of India. He regularly provides peer review for journals in his field (e.g., journals from IEEE, Kluwer and Elsevier), and has frequently been invited to review books published by Springer-Verlag and McGraw-Hill. In recognition of his teaching and research, he has been given the AICTE Career Award (1997-2000), the highest honor offered to young talented academicians by the All India council of Technical Education, Government of India.

Dr. Konar is a Principal Investigator or Co-Principal Investigator of four external projects funded by University Grants Commission (the UGC is one of the main federal funding agencies in India) and two projects funded by the All India Council of Technical Education, Government of India. The research areas of these projects include decision support system for criminal investigation, navigational planning for mobile robots, AI and image processing, neural net based dynamic channel allocation, human mood detection from facial expressions and DNA-string matching algorithms. Under his supervision, seven graduate students have already earned the degree of Ph.D., and three Ph.D. dissertations are in progress. Currently, he serves on the editorial board of the International Journal of Hybrid Intelligent Systems. Dr. Konar is the coordinator of the image processing part of the "Second Hooghly Bridge Project", one of the major projects of West Bengal Government.

Dr. Konar served as a member of Program Committee of several International Conferences and workshops, such as Intl. Conf. on Hybrid Intelligent Systems (HIS 2003), held in Adelaide, Australia and Int. Workshop on Distributed Computing (IWDC 2002), held in Calcutta.

Ajit K. Mandal holds an M. Tech Degree in Radio Physics and Electronics and earned a Ph.D. from the University of Calcutta, India. He is currently a Professor in the ETCE Department, Jadavpur University, Kolkata and served it as its Head from August 1992 to July 1994.

He was the chief investigator of a number of research projects funded by UGC, DST and AICTE and was the Scientist in Charge of the Eastern Regional Center of "Appropriate Automation Promotion Programme" funded by Dept of Electronics Govt. of India in collaboration with UNDP (1988-1990).

His teaching and research interests are Fuzzy Logic, Neural Networks, Evolutionary computing and Machine learning with applications to pattern recognition and digital image processing. He has about 36 years of research and teaching experience in the above areas. He has published numerous Technical papers in National and International Journals of repute and supervised numerous Ph. D. and Master of Engineering thesis. He has also authored a book " Introduction to Control Engineering – Modeling , Analysis and Design", New Age International (P) Ltd. New Delhi.

He has chaired sessions in number of National and International Conferences and delivered seminar lectures at reputed national and international institutes

Dr. Mandal is a senior member of IEEE (USA) and acted as the Chairperson of Computer Chapter of IEEE Calcutta Section from January 2001 to December 2004. He is a Fellow of the Institution of Engineers, India and Fellow of the Institution of Electronics and Telecommunication Engineers, India.